First Steps in Space-Time

First Steps in Space-Time: A Brief Introduction to Special Relativity provides an accessible, authentic, and readable introduction to the theory of special relativity. The academic level of the book builds only on skills that would be covered in a high-school maths course, such as Pythagoras's theorem, and rearranging equations, and no prior knowledge of relativity (or physics) is assumed. The key benefits of the work are to bridge the gap between popular science books and university textbooks, and make the theory of relativity as broadly accessible as physically possible. The book allows the reader to discover and appreciate that relativity is not an intractable esoteric curiosity, but a beautiful and succinct theory that profoundly shaped the course of history and our interpretation of our day-to-day lives. This book is ideal for readers with an interest in physics and a little working knowledge of maths, who have studied mathematics at high-school level: professionals such as accountants, bankers, doctors, economists, engineers, lawyers, pharmacists and teachers, or high school students who might be considering studying physics or a related subject at university.

Ed Macaulay is a lecturer in Physics and Data Science at Queen Mary University of London, where he's keen on making physics more accessible. His research background is in cosmology and dark energy, and he has interests in outreach and Physics Education Research. He is originally from London, and he has an MSci in Physics from Imperial College, London, and a DPhil in Astrophysics from the University of Oxford.

First Steps in Space-Time
A Brief Introduction to Special Relativity

Ed Macaulay

CRC Press
Taylor & Francis Group
Boca Raton London New York

CRC Press is an imprint of the
Taylor & Francis Group, an **informa** business

Designed cover image: Albert A. Michelson and Edward W. Morley. On the Relative Motion of the Earth and of the Luminiferous Ether. Sidereal Messenger, 6:306–310, November 1887

First edition published 2025
by CRC Press
2385 NW Executive Center Drive, Suite 320, Boca Raton FL 33431

and by CRC Press
4 Park Square, Milton Park, Abingdon, Oxon, OX14 4RN

CRC Press is an imprint of Taylor & Francis Group, LLC

© 2025 Ed Macaulay

Reasonable efforts have been made to publish reliable data and information, but the author and publisher cannot assume responsibility for the validity of all materials or the consequences of their use. The authors and publishers have attempted to trace the copyright holders of all material reproduced in this publication and apologize to copyright holders if permission to publish in this form has not been obtained. If any copyright material has not been acknowledged please write and let us know so we may rectify in any future reprint.

Except as permitted under U.S. Copyright Law, no part of this book may be reprinted, reproduced, transmitted, or utilized in any form by any electronic, mechanical, or other means, now known or hereafter invented, including photocopying, microfilming, and recording, or in any information storage or retrieval system, without written permission from the publishers.

For permission to photocopy or use material electronically from this work, access www.copyright.com or contact the Copyright Clearance Center, Inc. (CCC), 222 Rosewood Drive, Danvers, MA 01923, 978-750-8400. For works that are not available on CCC please contact mpkbookspermissions@tandf.co.uk

Trademark notice: Product or corporate names may be trademarks or registered trademarks and are used only for identification and explanation without intent to infringe.

ISBN: 978-1-032-91479-4 (hbk)
ISBN: 978-1-032-90838-0 (pbk)
ISBN: 978-1-003-56355-6 (ebk)

DOI: 10.1201/9781003563556

Typeset in Latin Modern font
by KnowledgeWorks Global Ltd.

Publisher's note: This book has been prepared from camera-ready copy provided by the authors.

Dedicated to Mum
Valentine Macaulay
1955 – 2024

Contents

List of Figures ... ix

Acknowledgements ... xi

Note on Illustrations ... xiii

Author Biography ... xv

CHAPTER 1 • Marvels in Store ... 1

CHAPTER 2 • Perfect Timing ... 11

CHAPTER 3 • Great Lengths ... 24

CHAPTER 4 • Made to Measure ... 31

CHAPTER 5 • The Fast and the Curious ... 38

CHAPTER 6 • Impulsive Reasoning ... 45

CHAPTER 7 • Work in Progress ... 55

CHAPTER 8 • A Momentous Integration ... 67

CHAPTER 9 • Time to Reflect ... 76

Bibliography ... 83

Index ... 85

List of Figures

1.1	The Michelson Interferometer	5
1.2	The 1919 Eclipse	9
2.1	Alice	14
2.2	Bob	15
2.3	The light clock	17
3.1	The Lorentz factor	29
4.1	Length contraction observed by Bob	32
4.2	Length contraction observed by Alice	34
6.1	A space-time diagram	47
6.2	A space-time diagram for an external observer	49
6.3	Comparing Newtonian and Relativistic momentum	53
7.1	A graph of constant force	57
7.2	A graph of constant velocity	58
7.3	A graph of increasing velocity	59
7.4	A graph of varying force	62
8.1	Comparing Newtonian and Relativistic energy	72

Acknowledgements

I would firstly like to extend my sincere thanks to the excellent staff at Taylor & Francis, including Rebecca Hodges-Davies, Danny Kielty, Shashi Kumar, Sanika Shar, and Natasha Skara and Riya Bhattacharya from KWGlobal. Their energy, expertise, and support have transformed this manuscript into the publication you are reading now.

I would also like to extend my thanks to my family for their encouragement, questions and support in the writing of this manuscript. I particularly appreciate the effort of my brother Jack Macaulay, who courageously read through the very earliest draft of this work, and whose feedback and encouragement has improved the book immensely.

I would also very much like to thank the staff and students at Queen Mary University London for their enthusiasm, energy, and encouragement while preparing this manuscript for publication. As is – of course – de rigueur: any oversights, omissions or mistakes left in this work remain entirely my own.

Note on Illustrations

Several of the figures that appear throughout this book are based on illustrations that have been made available as part of The British Library's Flickr Commons project of public domain illustrations from their collection. The drawing of the train carriage was adapted from an illustration from *'Souvenirs d'une mission aux États-Unis d'Amérique'* by Émile Malézieux. The book was published in Paris in 1874, just thirteen years before the famous 1887 Michelson–Morley experiment, which set the stage for relativity.

The illustrations for our two stalwart intrepid observers Alice & Bob are originally from illustrations by J.B. Partridge for *'The Works of G. J. Whyte-Melville'*, edited by Sir H. Maxwell. This book was published by W. Thacker & Co. in 1898; only seven years before Einstein's miraculous series of papers in 1905 that sparked the revolution of twentieth century physics. This text is also decorated with illustrations from *'The Half Hour Library of Travel, Nature and Science for young readers'*, originally published in 1896. These illustrations provide a little context as to the state of the art of astronomy while the seeds of relativity were being sown.

xiv ■ Note on Illustrations

MIDNIGHT SKY—MOON, STARS, AND COMET.

Author Biography

Ed Macaulay is a lecturer in Physics and Data Science at Queen Mary University London. He has a PhD in astrophysics from the University of Oxford, and a research background in supernova cosmology and dark energy. He loves teaching and outreach, and has taught courses ranging from introductory solar system astronomy to general relativity and cosmology. He's keen to make science more accessible, and has presented outreach events in cafes, pubs, cinemas, and even music festivals. He's presented outreach talks as part of events including Pint of Science, Astronomy on Tap, Stargazing Live, and Cafe Scientifique, and to audiences ranging from primary-school pupils to specialist astronomy groups. He's keen to make physics accessible, understandable, and engaging, and for everyone.

ABOVE THE CLOUDS, NIGHT.

CHAPTER 1

Marvels in Store

AT THE DAWN of the twentieth century, some of the most creative minds in Europe were developing new ways to represent our four dimensions of space and time. They developed new ways to depict the world, with counter-intuitive perspectives and paradoxical geometry. They even abandoned our traditional notions of simultaneity and the constant progression of time. Pioneers of the field included the likes of Jean Metzinger, Georges Braque, Albert Gleizes, and Pablo Picasso. Their movement became known as Cubism and had a profound influence on the art and culture of the twentieth century.

Before Cubism, the Impressionists had begun to experiment with different styles of painting, but there were still two hard and fast rules that artists had stuck to. Every painting – whether a still life, a portrait, or a landscape – was a snapshot of a single moment in time, and from the perspective of the artist. The Cubists realised that these two arbitrary rules are just that: arbitrary.

Whereas a traditional painter would prefer their subject to remain as still as possible, a Cubist painter might prefer their subject to dance around as they were painted. Our Cubist artist wouldn't just sit in the same spot for the whole painting. They'd move around their subject, including many varied angles and perspectives. Instead of just a snapshot of – say – a dancer, a Cubist painting might try and capture more of the essence of the dance and the dancer. However, if we don't understand Cubism, a Cubist painting like this can end up looking like a big confusing mess.

DOI: 10.1201/9781003563556-1

Cubism can certainly be challenging to our conventional notions of geometry, perspective, time, and space. However, as challenging as the paintings can be, if we only ever read about them, we would find them more confusing, not less so. We wouldn't attempt to understand Cubism without ever studying Cubist paintings. At some point, we might even wonder: What inspired the Cubists to take such a radical approach to their art?

Just a few years before Cubism, scientists were also developing radical new ways to think about time and space. The work was pioneered by scientists including Hendrik Lorentz, Henri Poincaré, and Albert Einstein, and the theory they developed was the Theory of Relativity. Relativity had a seismic effect on all of the twentieth century, transforming science, politics, culture, and even art.

Whereas the Cubists developed new ways to represent our four dimensions of space and time with oil paint on the surface of a two-dimensional canvas, the Relativists developed new ways to represent our four dimensions of space and time with mathematical equations. These equations provide a whole new perspective on time, space, and the universe we live in. To understand what the Relativists discovered, we have to understand their equations. Attempting to understand relativity without the equations is like attempting to understand Cubism without the paintings.

Although relativity can have a formidable reputation, it may come as a surprise that nearly all of the equations can be understood by starting just with mathematics from high school. There's no need to panic. Anyone who can recall their high school mathematics is ready to discover relativity for themselves.

Before relativity, the science of physics was generally regarded as a done deal. Newtonian mechanics, thermodynamics, and electromagnetism had all proven themselves sensationally successful, not just in the laboratory, but as the foundations of the industrial revolution. There were just a couple of slightly vexing questions left, and finding explanations to these quandaries was generally regarded as an exercise in dotting the 'i's and crossing the 't's. In

MOON APPROACHING THE SUN.

1894, the physicist Albert Michelson summed up the general mood pretty well:

'While it is never safe to affirm that the future of Physical Science has no marvels in store even more astonishing than those of the past, it seems probable that most of the grand underlying principles have been firmly established and that further advances are to be sought chiefly in the rigorous application of these principles to all the phenomena which come under our notice. It is here that the science of measurement shows its importance – where quantitative work is more to be desired than qualitative work. An eminent physicist remarked that the future truths of physical science are to be looked for in the sixth place of decimals.'

We'll hear more from Professor Michelson shortly. Four years later, in 1888, the astronomer Simon Newcombe put it more succinctly:

'We are probably nearing the limit of all we can know about astronomy.'

It's easy to see why Newcombe was so confident. At the time, one of the last thorny questions remaining in astronomy was the issue of an utterly minuscule discrepancy in the orbit of the planet Mercury. Astronomers had known since Johannes Kepler, two centuries earlier, that the planets orbit the sun in ellipses, not perfect circles. With decades of exquisitely precise observations, astronomers had discovered that these elliptical orbits themselves very slowly rotate about the sun, due to the gravitational influence of the other planets in the Solar System.

After painstakingly reviewing two centuries of measurements of the orbit of Mercury, Newcombe determined that its orbit was precessing about the Sun at a rate of 0.16 degrees per century. However, the subtle gravitational effects of the other planets could only account for a precession of 0.15 degrees per century, leaving an almost totally negligible difference of 0.01 degrees per century. It would take ten thousand years for the prediction to be off by even one degree. It would've been easy to simply shrug off such a tiny difference.

At the time, Michelson was absolutely right that physics appeared to have become the science of hunting for the sixth decimal place. Whilst he was correct to hedge his bets, he would've scarcely been able to imagine the marvels in store sparked by such an utterly minuscule difference. In fact, it was Michelson himself and his colleague Edward Morley whose experiment ignited the revolution of relativity.

Michelson & Morley were attempting to measure the effect of the motion of Earth on the speed of light [8]. In Figure 1.1, we can see a diagram from their 1887 paper, 'On the Relative Motion of the Earth and the Luminiferous Ether' of the apparatus they were using. What Michelson & Morley found perplexing was that the

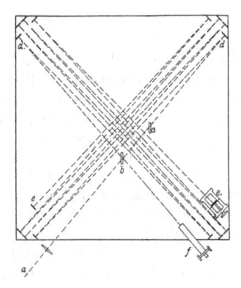

Figure 1.1: In this figure, we can see a diagram of the 'Michelson Interferometer' used by Michelson & Morley in their famous experiment [8]. For scale, the square block that the experiment was mounted on is about 1.5 metres across. The whole apparatus could be rotated so that one or the other of the arms could be set in the direction of motion of the planet Earth. Michelson & Morley hypothesised that the motion of the planet would cause the light to travel at a different speed in this arm than in the perpendicular arm. No difference was ever found, providing the first experimental evidence that the speed of light is constant.

motion of our planet appeared to have no effect whatsoever on the speed of light. Despite the rotation of the Earth about its axis and the motion of the Earth as it was orbiting the Sun, Michelson & Morley found that, as far as the beams of light in their experiment were concerned, it was as if the Earth was absolutely, perfectly stationary.

This result was rather troubling to physicists at the time, who would've expected light rays travelling in the direction of the motion of Earth to be faster than light rays travelling in a perpendicular direction. Driven by the Michelson & Morley experiment, physicists of the time began to develop creative ways to explain the result.

Woldemar Voigt and George Fitzgerald both suggested that this curious result could be explained if travelling at high speeds caused distances to be squashed along the direction of travel. Thinking along similar lines, Joseph Larmor proposed that high speed might cause the rate of the passage of time to slow down. These ideas were unified and pioneered by Hendrik Lorentz, who developed a complete model of how time and distance could be squashed to explain the Michelson & Morley experiment [7].

Voigt, Fitzgerald and Larmor were all thinking along the right lines, and the framework developed by Lorentz was mathematically perfect. However, there was a fundamental sticking point with their approach that none of these physicists could escape from. While it was certainly radical to propose that increasing speed might cause lengths to be squashed and time to pass more slowly, this speed and the corresponding effects on time and space were posited with respect to a hypothetical, universal, and absolute stationary frame of reference against which the speeds could be measured. The problem was, nobody could find a suitably static absolute frame of reference.

Planet Earth couldn't be this perfect, absolute frame of reference, because it's orbiting around the Sun. But the Solar System couldn't be a candidate either, because the Sun is itself orbiting about the Milky Way galaxy, and, as was discovered later, our Milky Way galaxy is moving with a stupendous speed relative to other galaxies. And yet none of this motion appeared to have the slightest effect on the Michelson & Morley experiment, which, to all intents and purposes, appeared to be perfectly stationary.

Amidst these astronomical quandaries, a brilliant cohort of physicists were studying the behaviours of electricity and magnetism. This generation included the likes of André-Marie Ampère, Georg Ohm, Carl Friedrich Gauss, Joseph Henry, and Michael Faraday, whose names are now immortalised in the units of electrical circuits. These physicists observed and described relationships between the interactions of magnetic fields and electrical current flowing through wires. These relationships were triumphantly combined into a single, unified theory of electromagnetism by the brilliant champion of classical physics, James Clerk Maxwell. Maxwell's unified theory made the curious prediction that light would always travel at a constant speed. This naturally prompted the question: if light always travels at a constant speed, what is this constant speed relative to?

This situation remained a perplexing mystery until 1905, when Albert Einstein proposed a new interpretation of the mathematical framework that Lorentz had developed [4]. In Lorentz's model, the time dilation and length contraction both depend on a speed, which Lorentz had presumed to be a speed relative to our elusive, absolute frame of reference.

Einstein's leap of perspective was to realise that no such absolute frame of reference exists, and the speed that should be considered in Lorentz's model is the relative speed between the observer and the experiment. The key word there is 'relative'. That's why it's called the Theory of 'Relativity'.

With Einstein's relativistic interpretation of Lorentz's model, Michelson & Morley's results are a natural and inexorable consequence. Michelson & Morley were stationary relative to their experiment, and that was all that mattered. This interpretation also answered the question of Maxwell's electromagnetism as to what the speed of light was constant relative to. In this interpretation, the speed of light is a constant relative to everything. In the next chapter, we'll see exactly what this means. As we'll see, Einstein's relativistic approach soon drew him to the conclusion that matter is another form of energy [3].

After 1905, the pace of the development of relativity accelerated dramatically, with major contributions by physicists including Hermann Minkowski, Henri Poincaré, and Max Planck, among many others. However, there was a significant limitation to relativity at the time. The theory that the Relativists were developing could only be applied in the special case of an object moving in a perfectly straight line. This theory came to be known as 'special relativity'. The mathematics developed by Lorentz couldn't be directly applied to the general case of objects moving on curved paths.

It may sound like a fairly modest endeavour to expand a theory from the special case of a straight line to include curved paths, but this generalisation proved tremendously more difficult. Whereas Einstein developed the essentials of special relativity in a matter of months, developing a general theory of relativity required a solid decade of work, based on the geometry of curved surfaces developed by the mathematician Bernhard Riemann.

By 1915, Einstein had developed the theory of general relativity, although he feared that his equations were so formidable that they might never be solved. He needn't have worried; less than a year later, the physicist Karl Schwarzschild found the first solution (while fighting in the trenches of World War One!).

With general relativity and Schwarzschild's solution, physicists were at long last able to calculate the dynamics of orbiting bodies, in particular, the planets of our Solar System. Amazingly, this radical new approach to gravity predicted that the orbit of the planet Mercury would precess about the Sun by an extra 0.01 degrees per century, compared to the Newtonian theory, and exactly as measured by Newcombe several decades earlier.

The landmark test of general relativity was not long coming. Schwarzschild's result predicted that our Sun was sufficiently massive to cause a very subtle deflection in the apparent position of distant stars. The only problem with this prediction is that any star close enough to the Sun to be deflected would be vastly outshone by the Sun itself. However, astronomers knew that in 1919 a total solar eclipse would be observable from Madagascar. At the totality of the eclipse, the Sun would be perfectly obscured by the Moon, allowing the position of the stars to be measured.

The astronomer Arthur Eddington led an intrepid expedition to the island to put the theory to the test [2]. One of the renowned photographs from the expedition is shown in Figure 1.2. Eddington's results were brilliantly summarised in a famous *New York Times* headline:

'LIGHTS ALL ASKEW IN THE HEAVENS;
Men of Science More or Less Agog Over Results of Eclipse Observations.

Figure 1.2: One of the photographs of the total solar eclipse from Eddington's 1919 expedition. By observing the deflection of light from stars close to the eclipsed Sun, the expedition provided astonishing confirmation of Einstein's theory of relativity. The photo is from the paper by Dyson, Eddington & Davidson,(1920) [2]. *'A Determination of the Deflection of Light by the Sun's Gravitational Field, from Observations Made at the Total Eclipse of May 29, 1919'*.

EINSTEIN THEORY TRIUMPHS'

The paper further provided a reassuring elucidation of the results:

'Stars Not Where They Seemed or Were Calculated to be, but Nobody Need Worry.'

Relativity had upended long-held traditional intuitions about the nature of space and time. Before relativity, time and space were regarded as an uninteresting blank canvas over which our universe was painted. The Relativists discovered that this canvas is itself a dynamic and animated entity, intrinsically interwoven with matter and energy. For more on the extraordinary story of the history of relativity, see, for example, references [1, 5, 6].

Although we're quite familiar with thinking about how much time and space things take, it can seem rather esoteric to think about time and space as intrinsic quantities themselves. It's important to remember that even Einstein had a great deal of help from many of his contemporaries in developing relativity. In this book, we're going to discover relativity by following the line of reasoning developed by Lorentz in order to explain the curious fact of the constant speed of light. With this framework, we can understand Einstein's key results about the relationships between time, space, matter and energy.

CHAPTER 2

Perfect Timing

THE SPEED OF LIGHT is exactly two hundred and ninety-nine million, seven-hundred and ninety two thousand, four hundred and fifty-eight metres per second. All of the extraordinary results from relativity stem from the fact that this speed is an absolute constant, everywhere, for everyone. But what exactly does it mean for the speed of light to be an *absolute* constant?

Let's start with a classic example: the situation of a passenger on a Train – Alice – as the train chugs past Alice's acquaintance – Bob – while he's waiting on a nearby train platform.

Let's suppose the train has a reasonable velocity for a locomotive, say, ninety miles per hour (or forty metres per second). Now let's suppose that Alice walks along the train carriage towards the front of the train. If Alice walks with a speed of two metres per second on the train carriage, what's her combined speed relative to Bob, who's waiting on the train platform? We can find this combined speed by adding together the speed of the train and Alice's walking speed. Let's write that out as a little equation:

$$40 \, \text{m}/\text{s} + 2 \, \text{m}/\text{s} = 42 \, \text{m}/\text{s}$$

So we find that the combined speed of Alice relative to Bob is forty-two metres per second.

Now let's suppose Alice stops walking and turns on the headlights in the front of the train. The photons are going to be moving away from her at exactly 299,792,458 metres per second. Meanwhile, Bob is still standing on the train platform and observing the train. Remember, the train is moving at 40 metres per second relative to him, and the photons are streaming out of the front of the train at 299,792,458 metres per second.

We might be sure to conclude that, relative to Bob, the photons must be travelling at 299,792,458 + 40, or 299,792,498 metres per second. Let's not be so hasty. What it means for the speed of light to be constant is that Bob will *also* see the photons travelling at 299,792,458 metres per second. On the face of it, this might appear to be an impossible paradox. How can Alice and Bob both measure the same speed of light?

We might suppose that, compared to the speed of light, forty metres per second is not much of a great deal. What if our train was a little faster? Let's imagine we sent our train down the track at half the speed of light. As before, Alice illuminates the light at the front of the train, and still observes the photons streaming ahead at the speed of light. Let's consider what Bob would observe while he's standing on the platform.

The train is hurtling past at half the speed of light. We might reasonably insist that Bob *must* now observe the photons streaming out at a grand total of one-and-a-half times the speed of light. As before, we shouldn't be so hasty. The essential empirical fact on which relativity is based is that Bob would *still* observe the photons travelling at exactly the speed of light, and not one iota more, despite the now decisively creditable speed of our locomotive.

That's what it means for the speed of light to be an absolute constant. It's not just constant if we're at rest relative to the source of the light. It's an absolute constant, for everyone, no matter how fast they're moving relative to the source. How can we resolve this counter-intuitive quandary?

By way of an analogy, it may help to imagine a family with just one impossibly difficult person, who always insists that they must have everything exactly their way. We might imagine that they only want to eat dinner at exactly quarter-past seven in the evening, so everyone else just has to work their schedules around that. If such a scenario does not, in fact, require a particularly taxing stretch of the imagination, then so much the better.

The speed of light is somewhat like that one impossibly difficult person. The speed of light will categorically, exclusively, only ever travel at 299,792,458 metres per second, relative to everyone. As such, everything else in the universe: time, space, mass and energy, just has to work around that. In a nutshell, the theory of relativity describes exactly how they do that so that the speed of light is always constant for everyone. To make that happen, the Relativists concluded that we would be required to jettison many of our intuitive notions about how time, space, mass and energy behave.

Over the next four chapters, we'll discover exactly how to weave space and time together so that the speed of light is always a constant. We'll then consider some of the extraordinary consequences of our newly stretched and squashed approach to spacetime. Let's begin with the first piece of the puzzle: time. How exactly does time contort itself so that we always have a constant speed of light?

In this classic example, Alice has set up a clock to measure the passage of time in the train carriage. Whereas a typical clock might measure the passage of time by counting the swings of a pendulum or the oscillations of a quartz crystal, Alice's clock marks the passage of time with pulses of light. Let's call it a 'light clock'. The light clock is illustrated in Figure 2.1.

The clock emits a pulse of light from the floor of the train, up to the ceiling. Let's call this one 'tick' of the light clock. How long will this take? Let's say that the ceiling is two metres above the floor. Remember, the speed of light is exactly 299,792,458 metres per second. How long would it take to cover those two metres?

We know that speed is distance divided by time. If we rearrange this equation a little, we can calculate the time by dividing the distance by the speed:

$$\frac{2.00 \, \text{m}}{299,792,458 \, \text{m/s}} = 6.67 \times 10^{-9} \, \text{s}.$$

so we find that each tick of the light clock would only take 6.67 nanoseconds. Can we generalise this a little? What if the height of the ceiling isn't exactly two metres? Let's say the height is just 'h' metres. How long would this take?

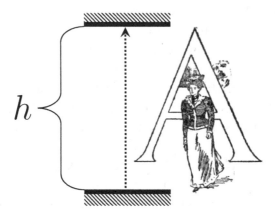

Figure 2.1: Alice observes a beam of light travelling a distance h.

We'll represent the speed of light with the letter c. Precisely why we use 'c' for 'speed of light' has been somewhat lost to time; it may be short for 'constant', or it may be from 'celitas', Latin for 'speed'. It might help to unceremoniously mangle both explanations together and remember that c represents a 'constant speed'.

Let's return our attention to the situation in hand. In general, how long will the light take for this trip? Let's call that time t'. The t' is clearly for 'time', but what about that little apostrophe? The apostrophe is called a 'prime', so we pronounce t' 'tee-prime'. We use a 'prime' whenever we are at rest relative to the quantity that we are measuring.

How we can relate that distance h to the speed of light, c, and the time, t'? Remember that a speed is a distance per time, so, solving for time, we find that

$$t' = \frac{h}{c}.$$

We can use this equation to calculate how long each tick would take, for any height of the ceiling. But what about Bob, watching

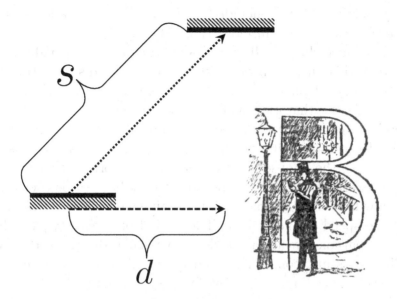

Figure 2.2: Bob is observing the ray of light from the platform. Since the train is moving relative to Bob, the train travels a distance d while the light ray is travelling. As such, Bob sees the light ray travel a greater distance, s.

the train speed by from the side of the platform? What's Bob going to see? How long will Bob observe each tick of the light clock lasting for?

Intuitively, we might say that, *of course*, Bob would say that each tick of the light clock takes exactly the same amount of time as for Alice. But let's think carefully about what Bob would observe. He's still going to see the light pulse going up from the floor of the train. But while this is happening, the whole contraption is moving along with the train carriage. So Bob will see the light pulse travelling a greater distance than Alice will. The situation for Bob is illustrated in Figure 2.2. If we wanted to impose our intuitive notion that events take the same length of time for everyone, we would have to impose that pulse of photons to travel faster than the speed of light, which is – of course – right out. So, if we want to work around the fact that the speed of light is the same for everyone, we have to conclude that Bob will observe a longer time for each tick of the light clock than Alice will.

How much slower will Bob observe each tick of the light clock? To work this out, we have to think about the distance travelled by the pulse of photons from Bob's point of view. Let's call the length of time that Bob will observe the tick to be t. Note that there's no prime on this t, because Bob isn't moving along on the train. Even though it's the same letter of the alphabet, this t is different to the t' observed by Alice. It's important that we keep them separate, and don't impose that they have to be the same.

How far along the track will the train go during this tick? Let's call the speed of the train v. The 'v' is from 'velocity', which in physics parlance means a speed in a particular direction. However, we'll always just consider motion in one direction, so we can use the terms 'speed' and 'velocity' fairly synonymously.

Let's remember that speed is a distance per time, so we can write down the distance that the train will travel (relative to Bob), during one tick of the light clock:

$$d = vt$$

So, from Bob's point of view, during one tick, the light pulse is going to cover a horizontal distance of d. However, the pulse still has to go that vertical height of h.

This brings us to the key picture to understand how we need to go about squashing time, so really do pay attention here: We can imagine the path that the light pulse will take as the hypotenuse of a right-angled triangle, where the base is d and the height is h. We can see a diagram of this hypothetical hypotenuse in Figure 2.3.

What will the total distance travelled by the photons be, from Bob's point of view? We can use Pythagoras's theorem to work that out, by calculating the length of the hypotenuse of our right-angled triangle. Let's call the hypotenuse s, so that we have:

$$s^2 = d^2 + h^2$$

What this allows us to do, with just a dash of algebra, is relate the time that Bob observes for one tick, t, to the time that Alice observes, t'. Remember that s is the distance that Bob observes

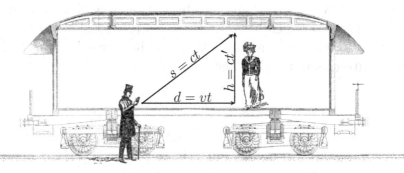

Figure 2.3: Alice (travelling on the train) observes a beam of light travelling a distance h in a time t'. Meanwhile, Bob (on the platform) observes the same beam of light travelling a greater distance, s, in a time t. Since Alice and Bob must both observe the beam of light travelling with the same speed, c, we have to conclude that Bob will observe this little event as taking more time than it does for Alice. If Bob observes the light-clock from the platform, he'll see the light travelling along the hypotenuse of this right-angled triangle.

the light pulse taking in his time, t. We know that the light pulse has to be travelling at the speed of light, c, so we know that we can relate this speed, time and distance:

$$s = ct.$$

We also have this distance as the hypotenuse of our right-angled triangle. Let's take the value $s = ct$ and substitute it where s appears as the hypotenuse:

$$(ct)^2 = d^2 + h^2$$

What next? How about that d there? Remember that this d is the base of our right-angled triangle, the horizontal distance that the train will travel relative to Bob in his time for one tick, t. We know that $d = vt$. Let's do the same thing as we did before with s, and substitute in our values for d:

$$(ct)^2 = (vt)^2 + h^2.$$

18 ■ First Steps in Space-Time

Now we only have one more substitution to go! Just that last height, h. Let's remember that we had a relationship between this height and the time for a tick that Alice would observe. Let's substitute this in for that last h:

$$(ct)^2 = (vt)^2 + (ct')^2.$$

Don't forget the important difference between t' and t. t' is the time that Alice observes, and t is the time that Bob observes. Now that we have them related, we're ready to compare them. Let's start by bringing both of the terms containing t over to the left-hand-side of the equation:

$$(ct)^2 - (vt)^2 = (ct')^2.$$

How can we simplify this a bit more? Let's start by dividing both sides of the equation by c^2:

$$t^2 - \left(\frac{v}{c}\right)^2 t^2 = t'^2.$$

And now we can factor out that t from both of the terms on the left-hand-side of the equation:

$$\left(1 - \left(\frac{v}{c}\right)^2\right) t^2 = t'^2.$$

Now we're ready to solve for Bob's time, t. Let's divide both sides of the equation by the term in parentheses, so that we have:

$$t^2 = \frac{t'^2}{\left(1 - \left(\frac{v}{c}\right)^2\right)}.$$

Finally, let's take the square-root of both sides of the equation:

$$t = \frac{t'}{\sqrt{1 - \left(\frac{v}{c}\right)^2}}.$$

The term in the denominator is tremendously important. This factor is used so often in relativity that it's given is own special symbol, 'gamma':

$$\boxed{\gamma = \frac{1}{\sqrt{1 - \left(\frac{v}{c}\right)^2}}.}$$

This is called the 'Lorentz Factor', after Hendrik Lorentz, who was the first to calculate it. The Lorentz Factor is so important because, as we'll see, it affects everything that we'll observe in relativity.

It's a very handy symbol, because it allows us to write the relationship between t and t' in a compact way:

$$\boxed{t = \gamma t'.}$$

What does this little equation mean? It may appear unassuming, but it is in fact amongst the most profound equations that humans have ever written.

Before this equation, time was considered an absolute quantity, whether it was or the lifetime of a muon, or the lifetime of a human. If an event lasts for, say, an hour for one person, then everybody else would also agree that the event lasts for exactly one hour. However, this equation tells us that time is not absolute. It tells us that the duration of time between two events for one person might be different if observed by someone else. We call this effect 'time dilation'.

While this may at first seem impossibly esoteric, our effort invested in working out the equation allows us to see exactly how this works with a concrete example. Let's suppose that one tick of Alice's light clock lasts one nanosecond. How long would this tick last if the train hurtles past Bob at two-thirds the speed of light? Let's remind ourselves exactly how t and t' are related:

$$t = \frac{t'}{\sqrt{1 - \left(\frac{v}{c}\right)^2}}.$$

We know that $t' = 1$ nanosecond. We know that the train is travelling at two-thirds the speed of light, so that fraction v/c will be equal to 2/3. Substituting these numbers in, we find:

$$t = \frac{1 \text{ nanosecond}}{\sqrt{1 - \left(\frac{2}{3}\right)^2}}.$$

Calculating these values, we see that the time that Bob will observe for one tick of the light clock will be 1.34 nanoseconds. In other words, when Bob observes the train, time will appear to be passing 34% slower than for Alice, on the train. While the notion that time is relative may still seem rather esoteric, hopefully even the most unwilling student of mathematics would agree that, while the calculations involved might not be entirely trivial, they are also by no means impossibly difficult.

Before marching on any further, let's take a moment to reflect on a couple of important points. It's most straightforward to interpret the results of this thought experiment just in terms of the rate of ticking of the light lock. But this doesn't only mean that ticks of light clocks would be running slowly. Ticks of mechanical clocks would be running slowly. Heartbeats would be running slowly. It's not the case that everything on the train carriage is running slow in a mechanical sense, like a metronome running down. From Bob's point of view, *everything* on the moving train carriage is running slowly, because *time itself* is running slowly. For Bob, on the platform, time is still passing at the usual rate, because time is not absolute.

The last point to note for now is that special relativity has no more to say has to why or how time is running slowly. It just states that, for the speed of light to be an absolute constant for everyone, time must pass more slowly when observing events in moving reference frames.

Let's now send the train past the station at a stupendously fast 90% of the speed of light. We can work out how much slower Bob will observe Alice by calculating the Lorentz factor. The train's now going at 90% of the speed of light, so that v/c fraction will be 0.9. So the Lorentz factor will look like:

$$\gamma = \frac{1}{\sqrt{1 - (0.9)^2}}.$$

On evaluating the numbers, we find that Bob will observe time passing on the train at a rate 2.3 times slower than it does for him.

Suppose Bob could see a clock on the wall of the train carriage. By the time he's seen the big hand on the clock move forward by one minute, two minutes and eighteen seconds will have passed for him. If Alice, say, took half an hour enjoying a game of solitaire, the game would appear to Bob to last an hour and nine minutes. To Bob, everything on the train carriage would be transpiring in slow motion.

What would this feel like for Alice? Would she notice that everything is happening on the train in slow motion? To Alice, everything on the train would appear exactly as usual. It's not just clocks and equipment that's running slowly on the train. It's time itself. It's everything, including Alice. So, as far as Alice is concerned, everything on the train is running exactly as usual. But suppose she looked out of the window, and saw Bob on the platform. What would events on the platform look like to Alice? It's here that things become really interesting.

We might intuitively conclude that if Alice appears to Bob to be running in 2.3 times slow motion, then Bob must appear to Alice to be running at a rate 2.3 times sped up. Let's calculate exactly what Alice would see, and test this hypothesis. Remember, we can calculate the relative rate of time by calculating the Lorentz factor. Here it is again:

$$\gamma = \frac{1}{\sqrt{1 - \left(\frac{v}{c}\right)^2}}.$$

What's going to change if we want to calculate the rate for Bob as observed by Alice, instead of, as previously, the rate for Alice as observed by Bob? The Lorentz factor only depends on the relative speed between the two observers, v. The train goes past Bob at 90% the speed of light, so $v/c = 0.9$. From Alice's point of view, she's stationary in the train carriage, and it's the train platform that's moving past her at 90% the speed of light. The only thing that's going to change for Alice will be the direction.

If, from Bob's point of view, the train moving past him corresponds to a positive speed, then, from Alice's point of view, the same situation corresponds to Bob moving past her with a negative speed. So the only difference will be that $v/c = -0.9$. What effect will that minus sign have on our Lorentz factor? Let's put it into the equation and find out:

$$\gamma = \frac{1}{\sqrt{1 - (-0.9)^2}}.$$

We do have the minus sign in there now, but it's just about to disappear because we always square the v/c fraction. But what does this mean? It means that Alice will have exactly the same Lorentz factor when she observes Bob as Bob has when he observes Alice. So, from Alice's point of view, for every second that elapses for Bob, 2.3 seconds will elapse for her. If Bob takes a day to complete a jigsaw puzzle, it will appear to Alice that Bob takes the whole weekend.

Is this another frustrating paradox? How can Alice and Bob *both* observe each other in slow motion? We might think that surely if one of them appears to be in slow motion, they must appear to the other as sped up? What's going on? We can glean a little intuition if we take Alice off the train and stand her on the platform, facing Bob.

Let's say that they are both 180 cm tall. Let's tip Bob backwards by 45 degrees. To Alice, Bob's projected height is now 153 cm. Of course, Alice knows that Bob's true height is still 180 cm. He hasn't shrunk, his height above the ground is just less because we've tipped him back. What would this situation look like for Bob? If Alice sees him as shorter, does that mean that Bob will see Alice as taller?

Of course, this is something about which we do have some useful intuition. We chose to tip Bob backwards, but the important point is that we have really just increased the angle between them by 45 degrees. Bob could equally well consider Alice as now tipped back by an extra 45 degrees. We know that Bob wouldn't view Alice as taller. We're happy with the notion that Alice's projected height would also appear squashed by exactly the same amount.

APPEARANCE OF SUN IN AN ECLIPSE.

As for the time intervals we have been thinking about in this chapter, the shortest time that we will observe for an event is the time we would observe if we were stationary relative to the event, the 'proper time'. If we are moving with a speed relative to the event, it will always appear to take more time to us. There's no speed we could choose that would make the event appear to go faster. In the next chapter, we'll discover the amazing consequences of relativity when we try to measure the lengths of objects when we're moving at relativistic speeds.

CHAPTER 3

Great Lengths

WE'VE JUST SEEN THAT, to keep the speed of light constant for everyone, time has to be a relative quantity. If Bob is waiting on the platform and observes Alice on the moving train, he'll see time passing more slowly on the train. If an event lasts an hour for Alice, the event may appear to Bob to last an hour and a half (for example). What if, instead of measuring a time duration on the train, Bob wanted to measure the length of the train; its distance from front to back? How could he go about doing this? First of all, he'd have to know how fast the train was travelling relative to him. So how could he go about measuring the speed of the train?

He could start by marking out a fixed distance on the track, say, ten metres, and time how long it takes the train to cover this distance. Let's say he plants a pair of flags to mark out this distance. He could start his stopwatch when the train passes the first flag and stop it when the train passes the end flag.

One detail Bob must be careful about is to start and stop his timer when the same part of the train passes each flag. For example, if he started the timer when the front of the train passes the first flag, and stopped the timer when the back of the train passes the last flag, then his time would be a little off. To have a correct measurement of the speed of the train, he has to note when the same part of the train passes each flag. He could choose the front of the train, or the end of the train, or a particular point in between, as long as he sticks with the same point.

As long as Bob sticks with a particular point, his measurement of the speed of the train won't depend on how long it is. It could be a locomotive without any carriages, or a mile-long freight train. As long as he chooses the same point as the train passes the start and end flags, he'll be OK. Let's say that Bob chooses the very front of the train.

He starts the stopwatch as the front of the train passes the first flag, and stops it again as soon as the front of the train passes the end flag. He checks the time on his stopwatch: 48 nanoseconds. The flags are ten metres apart, so we can calculate the speed:

$$\frac{10\,\mathrm{m}}{48 \times 10^{-9}\,\mathrm{s}} = 2.1 \times 10^8\,\mathrm{m/s}.$$

So we find that the train must be travelling at 70% of the speed of light.

Now on to step two: measuring the length of the train. We know that's it's travelling at 70% of the speed of light. How can Bob measure the length? This time, he just needs to wait by the side of the track, and start his stopwatch when the front of the train passes, and stop it just as soon as the end passes. How much time does Bob measure the train takes to pass him?

This time, Bob records a time of 272 nanoseconds. We can now calculate the length of the train by multiplying this time by the speed of the train:

$$\left(272 \times 10^{-9}\,\mathrm{s}\right) \times \left(2.1 \times 10^8\,\mathrm{m/s}\right) = 57\mathrm{m}$$

So we find that the train must be 57 metres long. Splendid! Bob's measured the length of the train! It's 57 metres long. Before we become too excited, let's consider for a moment what this situation looks like for Alice.

Alice and Bob will both agree on the speed of the train relative to the platform. Bob would say, of course, that the platform is stationary, and the train is moving relative to him. But Alice could equivalently say that it's the train that's at rest, and it's the platform (and everything else in the world) that's moving past her, at the same speed, but in the opposite direction. Both points of view are equally valid. But what would Alice observe if she looked out of the window and saw Bob timing the 272 nanoseconds for the train to pass?

Since Alice is moving at 70% of the speed of light relative to Bob, she would see everything Bob does as happening in slow motion, at a rate of 1.4 times slower than normal. So 272 nanoseconds for Bob would be 381 nanoseconds for Alice. From her point of view, it took 381 nanoseconds for the train to pass Bob. Since they both would agree that the train is travelling at 70% of the speed of light, how long would Alice figure the train is, using these two facts? Let's calculate the length according to Alice by multiplying the speed of the train by the time as observed by her:

$$\left(381 \times 10^{-9}\,\text{s}\right) \times \left(2.1 \times 10^8\,\text{m/s}\right) = 80\,\text{m}$$

So we find that Alice would say that the train is 80 metres long. Let's suppose that Alice already knows that the locomotive is 20 metres long, and so is each of the carriages. She has three carriages on her train, so, for her, the 80-metre length is exactly fine. But why then did Bob measure the total length of the train as 57 m? Is this another frustrating paradox?

The only way to resolve this quandary is to conclude that, like time, length must be another relative quantity. We can sort this situation out if lengths become contracted by exactly the same amount that time becomes slowed down. Let's recall that if an event takes a length of time t' for Alice moving on the train, Bob will observe a longer time, t:

$$t = \gamma t'.$$

And let's not forget that Alice will also see events running exactly as slowly when she observes Bob. We can calculate that little *gamma*, the Lorentz Factor, with our equation from the previous chapter:

$$\gamma = \frac{1}{\sqrt{1 - \left(\frac{v}{c}\right)^2}}.$$

The only way we can resolve our length measurement paradox is to conclude that lengths must also be contracted by γ. If Alice measures a length, L', Bob would observe a shorter length, L. We can relate these in a similar fashion as we do for time intervals:

$$\boxed{L = \frac{L'}{\gamma}.}$$

If Alice measures the length of the train to be 80 m, and she's travelling at 70% the speed of light relative to Bob, the Lorentz factor will be $\gamma = 1.4$, so Bob will observe a length of 57 m.

As with time, it's not just the train itself that Bob will observe to be squashed. Everything moving along the train will appear to be squashed (but only in the direction that the train is moving). This length contraction won't affect the width or height of the train (unless the train was also moving in those directions). This is why we were OK in the previous chapter to calculate the time dilation effect by considering photons travelling the height of the train carriage. Length contraction isn't going to affect this height.

We might worry about the base of our hypothetical right-angled triangle in our light-clock thought experiment: Might that be affected by length contraction? After all, it is in the same direction that the train is travelling? Fortunately, this distance isn't affected by length contraction, because it's not the length of anything moving on the train relative to Bob. It's the distance that a single fixed point on the train moves relative to Bob, due to the motion of the train.

If Bob observes the train being length-contracted as it chugs past him, what would all this be like for Alice on the train? Would she feel like she's being squashed? Would she notice that the seats don't have as much legroom as they used to? And what if she looks outside? If Bob sees her being squashed, does that mean the outside world will look stretched out to her?

Let's remember that Alice could equivalently say that she's the one at rest, and that it's everything else in the world that's moving past her. As far as she's concerned, she could be stationary, so she wouldn't notice any length contraction effects, just as she's not aware that, to Bob, she appears to affected by time dilation. What would the outside world look like to her?

As far as Alice is concerned, the train could be at rest, and we could repeat our thought experiment by moving the platform past her with the same speed, but opposite direction. If Alice says she's the one at rest, and the platform is moving towards her, then what would the platform look like? By the symmetry of the situation, we have to conclude that the platform must appear to have been squashed (along the same direction that it's moving towards us). We might ask about those flags that Bob carefully planted ten metres apart, what would they look like?

If the train is travelling at 70% of the speed of light, that ten metre distance will appear, to Alice, to be squashed down to 7.14 metres. So how come Alice still found that she was travelling at 70% of the speed of light, just like Bob did?

When Alice timed how long it took her to cover the distance between the two flags, it only took her 34 nanoseconds. 7.14 metres in 34 nanoseconds is still a speed of 70% of the speed of light. And what if Bob observed Alice timing how long it took to pass between the two flags? Remember that Bob observes everything on the train running at 1.4 times slow motion. So 34 nanoseconds for Alice is 48 nanoseconds for Bob, the exact time that he originally measured for the train to pass between the two flags he set ten metres apart. So Alice and Bob won't agree about the distance between the two flags, or how long the train takes to cover that distance, but they will both agree about the speed of the train relative to the flags.

If lengths and times are getting squished and squashed all over the place, how can we keep track of it all? The important thing is to measure the times and lengths when we're not moving relative to them. We call these the 'proper time' and the 'proper length'.

As we've seen, the Lorentz Factor is an essential quantity in relativity, affecting both time dilation and length contraction. We can see how the Lorentz Factor varies with speed in Figure 3.1. For speeds less than about 10% of the speed of light, the Lorentz

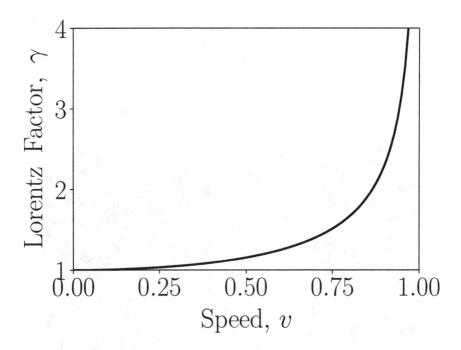

Figure 3.1: In this graph, we can see how the Lorentz factor, γ, depends on speed, v. The speed axis is relative to the speed of light, so a speed of, say, $v = 0.5$ would be half the speed of light. If someone is travelling with a speed of v relative to us, we will see their clocks running slow and their length squashed by a factor of γ. This effect starts out very small, and at v less than 0.1 it's almost negligible, but as our speed increases, it very quickly becomes a huge effect. As our speed approaches the speed of light, the Lorentz Factor shoots up to infinity!

Factor is almost exactly equal to one, so perhaps we shouldn't be surprised that all of this time dilation and length contraction is a bit counter-intuitive. In the next chapter, we'll discover some more extraordinary consequences of the Lorentz factor when Alice and Bob attempt to agree on the time and location of events.

MOON IN QUARTER.

CHAPTER 4

Made to Measure

IF ALICE AND BOB observe each other while Alice is riding the relativistic locomotive, we've seen how they will observe each other as affected by time dilation and length contraction. What if Alice and Bob attempted to coordinate an event with each other? How could they both agree on anything if space and time are getting squished and squashed all over the place? Let's see how they could do this with another thought experiment.

Alice is once again on track to zoom past Bob in the train at 70% of the speed of light. They've both remembered to bring along their stopwatches and have both agreed to start their clocks running at the instant that the back of the train passes Bob. Alice is sitting three metres from the back of the train. How far apart would Alice and Bob say they are from each other at the instant that they both start their stopwatches, when the back of the train is level with Bob?

For Alice, this is easy: She would just say 'three metres'. But what would Bob say? If the train is speeding past Bob at 70% of the speed of light, the Lorentz factor will be $\gamma = 1.4$. So any distance that Alice measures on the train will be squashed down for Bob by a factor of 1.4. If Alice sees the distance as 3 m, then Bob will see the same distance as only 2.1 m.

Let's see if we can generalise this. Next time, the train might be going at a different speed, and Alice might not be sitting exactly 3 m from the back of the train. Let's say that Alice would state that she's sitting x' metres from the back of the train. Remember

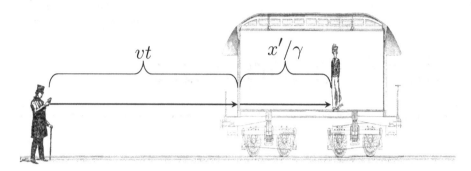

Figure 4.1: In this figure, we can see the train carriage, as observed by Bob. The train has been travelling at a speed of v for t seconds, so it's a distance of vt from Bob. Alice is a distance of x' along the train. Since the train is moving relative to Bob, it's going to appear squashed by a factor of γ.

that the prime symbol means that this is the distance that Alice would measure on the train. We can relate the distance that Alice is sitting from the back of the train train, x', to the distance that Bob would see, x:

$$\text{Bob sees}: x = \frac{x'}{\gamma}$$

But this is only true for the exact moment that the back of the train is lined up with Bob. What if Bob waits for, say, sixty nanoseconds? At 70% of the speed of light, the train would travel 12.6 m down the track in this time. So the back of the train is now 12.6 m away from Bob. And, from Bob's point of view, Alice is 2.1 m from the back of the train. So after 60 nanoseconds, Alice is now 14.7 m away, from Bob's point of view.

If Bob waits for a time of t, and the train is going at a speed of v, then the back of the train will be an extra distance of vt away from him. In Figure 4.1, we can see these two distances from Bob's point of view. Let's add this extra distance of vt onto our expression, so that it works for any time:

$$x = vt + \frac{x'}{\gamma}$$

Let's summarise what this tells us: If Alice says she's a distance of x' from the back of the train, and Bob waits for a time of t, then Bob will see that Alice is a distance of x from him. With a little bit of algebra, we can re-arrange this equation for x':

$$x' = \gamma\left(x - vt\right).$$

The reason that this dash of algebra was useful is that we now have everything that Bob could directly measure on the right-hand-side of the equation (the speed of the train v, some time interval for Bob, t, and how far away Bob sees Alice, x, at that point in time). We can now combine those thee values to find x': Alice's distance from the back of the train, from her point of view.

Can we come up with an equivalent relationship for Alice? Can we combine only the things that Alice could directly measure, and allow Alice to calculate how far away Bob sees her? Alice and Bob both agree on the speed of the train, v. Alice can directly measure x', how far away she is from the back of the train. She can also note the time on her stopwatch, t'.

So after a time for Alice of t', Alice would say that the train has travelled a distance of vt' down the track. For Alice, whichever seat she's sitting in, she would just say that she's x' ('x-prime') metres from the back of the train. So Alice would say that she's a distance of $vt'+x'$ away from Bob. As before, we can see these two parts of the distance in Figure 4.2, this time from Alice's point of view.

Remember, Alice could fairly say that she's the one at rest, and it's Bob who's moving away from her. So that $vt'+x'$ distance for Alice will be equal to the distance as seen by Bob, squashed down by γ. So, for Alice:

$$vt' + x' = \frac{x}{\gamma}$$

With just a little bit of algebra, Alice can work out how far away she would appear to Bob:

$$x = \gamma\left(x' + vt'\right)$$

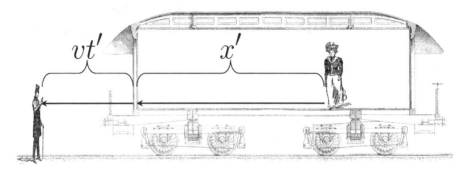

Figure 4.2: In this figure, we can see the same situation as before, but now from Alice's point of view. Alice is still a distance of x' along the train, but since she isn't moving relative to the train, Alice doesn't see the train squashed like Bob does. From Alice's point of view, she's stationary, and Bob is moving away from her, with a speed of of v, so the train is a distance of vt' away from Bob. t' is the time measured from Alice's point of view.

Let's write this equation down together with the equivalent equation from Bob's point of view:

$$\begin{cases} x = \gamma\left(x' + vt'\right) \\ x' = \gamma\left(x - vt\right) \end{cases}$$

The first equation is called the 'Lorentz Transformation' for position, and the second equation is called the 'Inverse Lorentz Transformation' (again, for position). The large squiggly brace connecting both equations just means that they are both valid simultaneously; we call them 'simultaneous equations'. If Alice is a distance of x' from the back of the train at a time of t' after she's started her stopwatch, Bob will see Alice at distance of x. If Bob sees Alice at a distance of x after he's waited for t seconds, then Bob can work out that Alice must be at a distance of x' from the back of the train.

Suppose Alice has been on the train for an arduously long time, say, $t' = 90$ nanoseconds. How could Alice work out how much time has passed for Bob, t, waiting at the station? The only relationship that Alice has between anything she can measure directly and that time for Bob (t) is the inverse Lorentz Transformation for position:

$$x' = \gamma(x - vt).$$

Alice certainly knows x', her position from the back of the train. And she knows v, too. But she couldn't solve for t unless she knows x, how far away she appears to Bob. But then she remembers that she can use the Lorentz Transformation for position to take her measurements of x' and t' and work out how far she appears to Bob, x:

$$x = \gamma(x' + vt').$$

So Alice could first work out x, and then use this value for x in the inverse Lorentz transformation, to, finally, solve for t. We can actually make this easier for Alice by doing doing the substitution and working through the algebra first. Let's take the expression for x from the Lorentz transformation, and substitute it in where x appears in the inverse Lorentz transformation. Let's take a big deep breath, because this starts out as a rather long equation:

$$x' = \gamma([\gamma(x' + vt')] - vt)$$

This might look like a big scary mess, but we've just made a substitution for x in the square brackets. We can simplify this equation considerably with a little algebra[1]. After solving for t, we find that

$$t = \gamma\left(t' + \frac{vx'}{c^2}\right).$$

This is called the 'Lorentz transformation for time'. It allows Alice to take the time on her stopwatch, t' and her position on the train, x', and work out what time Bob would see on his

---

[1]To work though the algebra, we need the relationship that:

$$1 - \frac{1}{\gamma^2} = \frac{v^2}{c^2}$$

stopwatch, t. Let's write the Lorentz transformations for time and position out together:

$$\begin{cases} t = \gamma \left(t' + \frac{vx'}{c^2}\right) \\ x = \gamma \left(x' + vt'\right) \end{cases}$$

Taken together, we can just call them the Lorentz transformations. They allow Alice to take her coordinates (her values of t' and x') and work out the corresponding coordinates for Bob (t and x).

What about the height of the train, or the width? Remember that length contraction only has an effect along the direction of motion, so if Alice says that the train is 3 m wide and 4 m high, then Bob will also say that the train is 3 m wide and 4 m high. Whatever direction the train is going, we can make our lives easier by choosing that direction as the x-axis, so that the y and z axes aren't affected. For completeness, we could write out the Lorentz transformations including the y and z coordinates, just to make it clear that they're the same:

$$\begin{cases} t = \gamma \left(t' + \frac{vx'}{c^2}\right) \\ x = \gamma \left(x' + vt'\right) \\ y = y' \\ z = z' \end{cases}$$

Since the y and z coordinates are both the same, we'll often just focus on the t and x coordinates.

This might all seem like a lot of work, just so that Alice and Bob can compare when and where events occur. We might be tempted to tell them just to be a little patient, and compare notes once the train has come to a stop. However, as we'll see, this ability that we now have, to relate coordinates as observed by Alice to those as observed by Bob (and vice versa), forms the bedrock of relativity. For now, we can only relate time and position, but we can combine these two fundamental building blocks to relate any quantities between our moving train and our stationary platform. As we'll soon see, it's this approach which leads to some of the most fascinating conclusions of relativity.

VOLCANIC ERUPTION IN SUN.

CHAPTER 5

The Fast and the Curious

WE'VE SEEN HOW ALICE AND BOB can relate coordinates between themselves. To keep the speed of light constant for everyone, time dilation and length contraction makes this business a little less straightforward than it would otherwise be. But what if Alice decides to go for a walk along the train. How fast would she appear to be moving, relative to Bob?

Let's start with an example with a realistic walking pace and a realistic train speed. Let's say that Alice is walking at three metres per second, and the train is moving at fifty metres per second. How fast would Alice appear to be moving relative to Bob? We could probably just figure out that she must be moving at a relative speed of fifty-three metres per second, but let's write it out in full anyway:

$$53 \,\mathrm{m/s} = 3 \,\mathrm{m/s} + 50 \,\mathrm{m/s}.$$

In general, if the train is moving with a speed of v, we can write this out as

$$u = u' + v$$

We've followed the convention of keeping quantities on the moving train 'primed' (with a dash), so u' is Alice's speed relative to the train (say, 3 m/s), and u is Alice's speed relative to Bob (53 m/s). This is the kind of relationship that might make comfortable, intuitive sense to us, and was one of the extraordinarily many contributions to science by Galileo Galilei. However, there's a couple of issues with the relationship if we think about speeds close to the speed of light.

Firstly, this relationship doesn't yield a constant speed of light for everyone. If Alice shines a laser in the direction the train is travelling, we could say that the speed of the laser beam is $u' = c$, and then Bob would see the photons travelling at a speed of $u = c + v$. One of the core ideas of relativity is that the speed of light is an absolute constant for everyone, so we need a way to combine two velocities like this so that Bob would observe the photons travelling at a speed of just $u = c$ (and not $u = c + v$). This is one case where our intuitive instincts don't sit well with nature at relativistic speeds.

The second issue with this standard, intuitive relationship, is that it would quite easily allow Alice to move at a speed greater than the speed of light relative to Bob. If the train is travelling at 70% of the speed of light, Alice would 'only' need to move with a speed greater than 30% of the speed of light to move relative to Bob at faster than the speed of light. We might contend that neither speed would be practical, but it would be theoretically possible. Another foundation of relativity is that nothing can move faster than the speed of light relative to anything else.

However, we do know that the equation $u = u' + v$ works just fine for everyday speeds. Any method that we come up with to combine our two speeds has to be able to recover this relationship at slow speeds.

Let's start by thinking about Alice's speed as she's walking along the train carriage, u'. What exactly do we mean if we say that Alice's speed is u'? What we mean is that she's changed her position (relative to the train) by some amount, $\Delta x'$. And she's changed her position in some amount of time, $\Delta t'$. If we divide this distance by this time, that gives us Alice's speed relative to the train:

$$u' = \frac{\Delta x'}{\Delta t'}$$

If Alice changes her position in a given time, how will this affect the position and time as observed by Bob? Remember that we can relate Alice's time and location relative to the train to her coordinates as observed by Bob by using the Lorentz transformations:

$$\begin{cases} t = \gamma\left(t' + \frac{vx'}{c^2}\right) \\ x = \gamma\left(x' + vt'\right) \end{cases}$$

We can find how Bob's values of t and x will change (Δt and Δx) by subtracting the Lorentz transformations at Alice's final coordinates ($t'+\Delta t'$ and $x'+\Delta x$) from their equivalents at Alice's original coordinates (which are just t' and x'). We need to take another deep breath, because it can look like a bit of a big mess. But remember, we're just subtracting one set of coordinates from the other:

$$\begin{cases} \Delta t = \gamma\left((t' + \Delta t') + \frac{v(x'+\Delta x')}{c^2}\right) - \gamma\left(t' + \frac{vx'}{c^2}\right) \\ \Delta x = \gamma\left((x' + \Delta x') + v(t' + \Delta t')\right) - \gamma\left(x' + vt'\right) \end{cases}$$

It might look like a big mess, but if we look a little closely, we'll see there's lots of nice cancellations that are going to happen. Once we've done all the cancelling out, the result looks much nicer:

$$\begin{cases} \Delta t = \gamma\left(\Delta t' + \frac{v\Delta x'}{c^2}\right) \\ \Delta x = \gamma\left(\Delta x' + v\Delta t'\right) \end{cases}$$

This is just the original Lorentz transformations, but for *changes* in coordinates, instead of absolute coordinates. Alice's speed relative to Bob is just the amount that she's changed her position from Bob's point of view (Δx), divided by the time it took her, again from Bob's point of view (Δt). This is how we can work out Alice's speed from Bob's point of view, u:

$$u = \frac{\Delta x}{\Delta t}$$

And, we've just worked out what Δx and Δt are, in terms of Alice's $\Delta x'$ and $\Delta t'$. So now we're all good and ready to work out u:

$$u = \frac{\gamma\left(\Delta x' + v\Delta t'\right)}{\gamma\left(\Delta t' + \frac{v\Delta x'}{c^2}\right)}$$

Again, this looks a bit messy, but we can probably spot that it might simplify quite nicely. Firstly, those two γ factors are just going to cancel out

$$u = \frac{\Delta x' + v\Delta t'}{\Delta t' + \frac{v\Delta x'}{c^2}}$$

And we have a $\Delta t'$ on the top and bottom of the fraction. What if we divided the top and bottom of the fraction by $\Delta t'$

$$u = \frac{\frac{\Delta x'}{\Delta t'} + v}{1 + \frac{v\Delta x'}{c^2 \Delta t'}}$$

but what's $\Delta x' / \Delta t'$? This is just Alice's speed relative to the train, u'. Once we substitute this in, we end up with a very important result:

$$\boxed{u = \frac{u' + v}{1 + \frac{u'v}{c^2}}}$$

What does this equation tell us? It gives us a way to relate Alice's speed relative to the train, u', to Alice's speed relative to Bob, u. Does this relationship fulfil our three requirements for a method to combine two velocities? Let's check.

What if Alice shines a laser beam down the train? The speed of the laser beam relative to the train will certainly be c, so we can say (in this example) $u' = c$. But what will Bob see? Let's work it out. If $u' = c$, then the speed of the photons as observed by Bob will be

$$u = \frac{c + v}{1 + \frac{cv}{c^2}}$$

A couple of those cs in the denominator will cancel to give us:

$$u = \frac{c + v}{1 + \frac{v}{c}}$$

And if we factor out a c from the numerator, we have:

$$u = c\frac{1 + \frac{v}{c}}{1 + \frac{v}{c}}$$

so $u = c$, and Bob will always observe the laser moving with a speed of c relative to him. It might seem like an impossible paradox to have the photons moving at the speed of light relative to Alice and also moving at the speed of light relative to Bob, but by thinking carefully about how time and distance are related for both of them, we can make it work.

What about the universal speed limit? Suppose that the train is travelling at $v = 0.7\ c$, and Alice is moving with a speed of $u' = 0.4\ c$ relative to the train. Our equation from the start would tell us that Bob would observe Alice moving relative to him with a speed 10% greater than the speed of light. Let's see how fast Bob would actually observe Alice:

$$u = \frac{0.4c + 0.7c}{1 + \frac{0.4c \times 0.7c}{c^2}}$$

working through the mathematics a little more, we find

$$u = \frac{1.1c}{1.28}$$

The numerator gives us the old result, but the denominator scales this down, so that Bob will observe Alice moving with a speed of $u = 0.86\ c$; faster than either Alice or the train individually, but still less than the speed of light.

It's somewhat of a situation of 'diminishing returns'. Remember that the faster the train moves relative to Bob, the more Bob will observe the train to be squashed by length contraction. So even if Alice covers the whole length of the train very quickly, if the train is already moving very fast, Bob won't observe Alice's extra speed as amounting to all that much.

Finally, what about our original example, where both the train and Alice were moving with more reasonable speeds for trains and humans:

$$u = \frac{3\,\mathrm{m/s} + 50\,\mathrm{m/s}}{1 + \frac{(3\,\mathrm{m/s}\, \times\, 50\,\mathrm{m/s})}{c^2}}$$

What matters for the scaling is how the product of the two speeds compares to c^2. For any everyday speed, this ratio is absolutely tiny, so we end up with

$$u = \frac{3\,\mathrm{m/s} + 50\,\mathrm{m/s}}{1 + 0.000...}$$

so Bob would still observe Alice moving along at 53 metres per second. An essential aspect of this equation is that, even if relativity predicts that counter-intuitive things will happen as we approach the speed of light, the theory must still reproduces the familiar, tested results when we return to more pedestrian speeds.

From the outset, it may seem like a paradox to live in a world where 50 m/s + 3 m/s = 53 m/s, and also where $0.4\,c + 0.7\,c = 0.86\,c$ (and not $1.1\,c$). When it comes to speeds like these (which are a good fraction of the speed of light), we really can't comprehend them on any kind of relatable human scale.

However, as slow as we might be, as humans we have a unique perspective: we can use equations. We've taken the constant speed of light as our starting point, and ventured out to explore the consequences, finding our way not with intuition, but with equations. We've arrived at an equation which challenges our conventional intuition about how speeds combine. This equation tells us that the speed of light will always be observed at the same value, and that we can never combine two speeds together to exceed the speed of light. This might still seem counter-intuitive, but, by charting a course set with mathematics, we can see how we arrive at these conclusions.

44 ■ First Steps in Space-Time

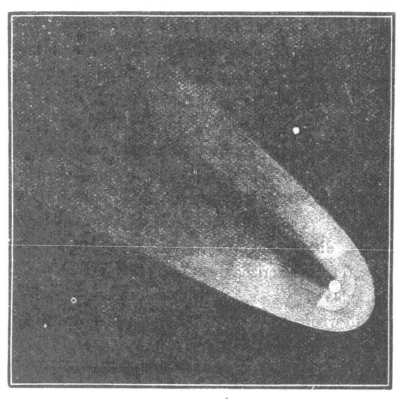

HEAD OF DONATI'S COMET.

CHAPTER 6

Impulsive Reasoning

ONE OF THE MOST IMPORTANT LESSONS from relativity is that the same event might not look the same to everyone who observes it. To keep the speed of light constant for everyone, the time duration of an event might be dilated, or the length of an object might appear contracted. If someone is moving with respect to something else, the combined speed will depend on the speed of the observer, and not in the straightforward way we might first imagine. No observer is right or wrong, just observing the same situation from a different point of view. One of Einstein's most important realisations was that, to keep the speed of light constant for everyone, relativity wouldn't just affect how objects and events appear, it would have to directly affect how objects behave and interact.

The most fundamental interaction we can imagine might be simply described as one object interacting with another object. That could be two objects bouncing off each other, sticking together, or splitting apart. It could be two cars crashing together, a proton and a nucleus splitting apart, or two billiard balls bouncing off each other. Centuries before Einstein, Isaac Newton realised that, whatever the type of interaction between two objects (whether that's two objects bouncing off each other, sticking together, or splitting apart), something will always be the same before and after the interaction.

The 'something' which Newton identified is called 'momentum', which – for historical reasons – is usually given the letter p. We can now write out the idea that 'the momentum before an interaction is equal to the momentum after an interaction' as an equation:

$$p_{\text{Before}} = p_{\text{After}}$$

Throughout this chapter, we'll follow the convention of keeping the 'momentum before' on the left-hand-side of the equation, and the 'momentum after' on the right-hand-side. What does this mean in practice? Let's try another thought experiment and see.

This time, Alice and Bob are both in train carriages. Both trains have exactly the same mass, which we'll call 'm'. The train carriages are coupled together and currently just sitting by the train station. Since neither train carriage is moving, the total momentum before must be zero:

$$p_{\text{Before}} = 0$$

However, in this scenario, we have a big coiled spring between the two carriages. When we uncouple the trains, the spring uncoils, and sends Alice and Bob in opposite directions down the track. Let's say that Alice goes to the left, and Bob goes to the right. In Figure 6.1, we can see an illustration of this situation.

After they have split apart, both Alice and Bob's trains are moving, so both have some momentum. The total momentum after we've uncoupled the trains is now Alice's momentum, plus Bob's momentum:

$$p_{\text{After}} = p_{\text{Alice}} + p_{\text{Bob}}$$

But we know that the total momentum must be the same before and after the two carriages spring apart, so we can write that as:

$$0 = p_{\text{Alice}} + p_{\text{Bob}}$$

In our example, Bob's mass is equal to Alice's mass. Even if we're not sure exactly what momentum is, if we only know that it depends on mass and speed, and we know that Alice and Bob's masses are equal, then we can conclude that Alice's speed away from Bob must be equal to Bob's speed away from Alice (in the opposite direction). If Bob is moving away to the right of our train

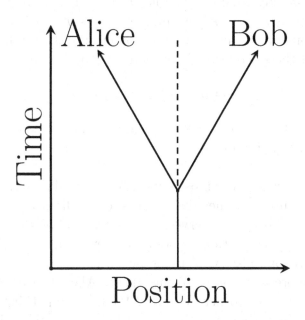

Figure 6.1: In this figure, we have a graph of Alice and Bob's position (on the horizontal axis), at different times, (on the vertical axis). Alice and Bob start off together. Because they are stationary, their position doesn't change as time increases. Once Alice and Bob split apart, Alice moves off to the left, and Bob moves off to the right, with the same speed as Alice, but in the opposite direction. The dashed line shows where Alice and Bob would've been if they hadn't split apart (this is called their 'centre of momentum'). From this point of view, the 'momentum before' is zero. Because Alice and Bob both have the same mass, and their speeds are equal and in opposite directions, the 'momentum after' is also zero.

line, let's say that his velocity relative to Alice is $+u'$. In that case, Alice's velocity relative to Bob must be $-u'$. What about their momentum? We know their momentum before they've split is zero. Is their combined momentum going to be zero now that they're both moving?

When Isaac Newton first introduced the concept of momentum, he defined it as the product of an object's mass and its velocity: $p = mv$. In our example, that means that Bob's momentum is $m(+u')$ and Alice's momentum is $m(-u')$. Let's write out our 'momentum before equals momentum after' relationship, and check if these values will conserve the total momentum:

$$0 = m(-u') + m(u') \checkmark$$

This is all well and good from the point of view of Alice and Bob. The 'momentum before' was zero, and the 'momentum after' is zero.

However, perhaps the most important tenet of relativity is that any law of nature (whether that's 'the speed of light is constant' or 'momentum is always conserved', or anything else) must be valid not just from one particular point of view, but from the point of view of anyone else who might be observing. What if, in our scenario, before the trains spring apart, Alice & Bob are both moving relative to another observer? We'll call our third observer Charlie. As with Alice and Bob, this choice is motivated only by alphabetical convenience.

Let's say that Alice & Bob are both moving with a speed of v relative to Charlie. We could consider that Charlie is at rest on the train platform, and that Alice & Bob's combined train is chugging to the right with a velocity of $+v$. Equivalently, we could consider that Alice & Bob's train is stationary on the platform, and that Charlie's train is chugging away to the left of the track. Both points of view are equivalent. We can see an illustration of this situation from Charlie's point of view in Figure 6.2.

The important thing is that Alice & Bob's momentum has to be the same before and after they split apart, both from their point of view, and also from Charlie's point of view. From Alice & Bob's point of view, it's a bit simpler, because, as far as they're concerned, their momentum is zero before and after. But what about from Charlie's point of view? Let's start by thinking about what the 'momentum before' is, from Charlie's point of view.

From Charlie's point of view, the velocity of Alice & Bob's train is v. But what about the mass? The mass of Alice's carriage is m, and the mass of Bob's carriage is also m, so the total mass of their train, before they split, is $2m$. So, from Charlie's point of

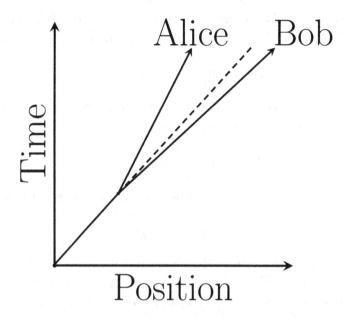

Figure 6.2: This is the same as the previous figure, but now from a point of view where Alice and Bob are already moving relative to an outside observer before they split apart. As before, the dashed line shows where Alice and Bob's trajectory would've been if they hadn't split apart. As with the previous figure, after the split, Alice moves off to the left of this trajectory, and Bob moves off to the right. However, because of the way that speeds combine in relativity, Bob's change in speed appears to be less than Alice's change in speed. If we use Newton's formula for momentum, it looks like there's less momentum after the split than before.

view, Alice & Bob's 'momentum before' isn't zero. If we say that momentum is mass times velocity, their momentum is $2mv$.

And what about the 'momentum after'? Alice & Bob have now split into their two separate train carriages, each with a mass of m. But how fast is Alice moving now relative to Charlie? And how fast is Bob moving relative to Charlie? They were both originally moving with a velocity of v. After springing apart, let's say that

Bob has increased his velocity by u', and Alice has decreased her velocity by u'. Does this conserve momentum? If momentum is mass times velocity, we might be tempted to write down:

$$2mv = m(v - u') + m(v + u').$$

Does this balance the 'before' and 'after' sides of the equation? Let's factor through the m:

$$2mv = mv - mu' + mv + mu'$$

and now combine both of those mv terms. Are both sides still equal?

$$2mv = 2mv + mu' - mu' \checkmark$$

That mathematics looks OK. Does that mean we're OK if momentum is $p = mv$? There's something important that we're missing.

If Bob is moving with a velocity of v relative to Charlie, and boosts his velocity by u', his new velocity relative to Charlie isn't just $v + u'$. We saw in the previous chapter that, when we combine speeds like this, to keep the speed of light consistent for everyone, we have to use this equation:

$$u = \frac{v + u'}{1 + \frac{u'v}{c^2}}$$

To find Bob's new velocity relative to Charlie, we have to divide that 'intuitive' combined velocity of $(v + u')$ by $(1 + vu'/c^2)$. Bob's velocity relative to Charlie after the boost is going to be:

$$u_{\text{Bob}} = \frac{v + u'}{1 + \frac{vu'}{c^2}}.$$

And what about Alice? Her velocity relative to Charlie won't just be $v - u'$. It's going to be:

$$u_{\text{Alice}} = \frac{v - u'}{1 - \frac{vu'}{c^2}}.$$

What effect is this going to have on momentum conservation? The key point here is that, from Charlie's point of view, Bob's change in speed is less than Alice's change in speed. If momentum does equal $p = mv$, from Charlie's point of view, it would look like there's less total momentum after separation than before.

If momentum does equal $p = mv$, we can only conserve momentum if we combine velocities in a way that doesn't give a constant speed of light for everyone. If we want to combine velocities in a way that does give a constant speed of light for everyone, and we want momentum to be conserved for everyone, then there must be more to momentum that just $p = mv$. What else could momentum be?

Before Alice & Bob's carriages split, Charlie would observe Alice & Bob's time running slow by a factor of γ. After splitting apart, Charlie would observe Alice's time running a little less slowly, and Bob's time running a little more slowly. What if their momentum also scaled by this way, too? What if momentum is:

$$p = \gamma m v.$$

What would our momentum conservation equation look like then? In our scenario, our 'momentum before' would be:

$$p_{\text{Before}} = \gamma_v\, 2m\, v,$$

and our 'momentum after' would be:

$$p_{\text{After}} = \gamma_{u_{\text{Alice}}}\, m\, u_{\text{Alice}} + \gamma_{u_{\text{Bob}}}\, m\, u_{\text{Bob}}.$$

Setting these two equal, we have:

$$\gamma_v\, 2m\, v = \gamma_{u_{\text{Alice}}}\, m\, u_{\text{Alice}} + \gamma_{u_{\text{Bob}}}\, m\, u_{\text{Bob}}$$

Don't forget that the three γ factors that appear here are all different, because they all depend on their corresponding velocities (the original speed, v, Alice's new speed, uAlice, and Bob's new speed, uBob). Is this the correct expression for relativistic momentum? We can make this a little easier to test by dividing out the masses from both sides of the equation:

$$2\gamma_v\, v = \gamma_{u_{\text{Alice}}}\, u_{\text{Alice}} + \gamma_{u_{\text{Bob}}}\, u_{\text{Bob}}$$

Let's test this out with an example. Let's suppose Alice & Bob are initially moving with a speed relative to Charlie of $v = 0.6\ c$. After they spring apart, they change their speeds by $u' = 0.2\ c$. How fast will Alice and Bob be moving relative to Charlie?

Using our relativistic velocity addition, we can calculate that for Alice, $u = 0.45\ c$, and for Bob, $u = 0.71\ c$. We can see here that, relative to Charlie, Bob has changed his velocity by less than Alice has. If $p = mv$, some momentum would have just evaporated from the universe.

Now that we have v and u for Alice and Bob, let's put these values into our momentum conservation equation:

$$2\gamma_v\, 0.6c = \gamma_{u_{\text{Alice}}}\, 0.45c + \gamma_{u_{\text{Bob}}}\, 0.71c$$

Now we just need to calculate the Lorentz factors for each of those velocities. For $0.6c$, $\gamma = 1.25$. For $0.45c$, $\gamma = 1.12$, and for $0.71c$, $\gamma = 1.42$. Let's put those numbers in, and see if the two sides of the equation are equal:

$$2 \times 1.25 \times 0.6c = 1.12 \times 0.45c + 1.42 \times 0.71c$$

Let's check the mathematics:

$$1.5 = 0.5 + 1.0\ \checkmark$$

So if we combine velocities relativistically, the conserved quantity isn't $p = mv$, it's:

$$\boxed{p = \gamma mv}$$

This quantity is called the 'relativistic momentum', in contrast to the original 'Newtonian' momentum, first proposed by Newton. We can see how both expressions vary with speed in Figure 6.3.

How does this extra factor change how objects behave? The key difference is that, in Newton's model, momentum just increases steadily as we go to higher speeds. A finite amount of momentum can take us to light speed, and there's no issue going even faster. With the relativistic model, momentum shoots up dramatically as we go to higher speeds. At about 87% of the speed of light, an object has twice the momentum that we would expect with

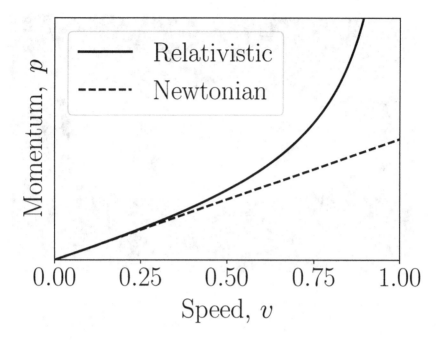

Figure 6.3: In this plot, we can see how momentum, p depends on speed, v. In the 'Newtonian' model, where $p = mv$, momentum just increases directly with speed. For speeds less than about $v = 0.2$, we can hardly see the difference between the Newtonian and the relativistic momentum, where $p = \gamma m v$. As we go to higher speeds, that γ has a huge effect!

Newton's equation. As we keep pressing still faster, the momentum shoots up towards infinity! We would need an infinite amount of momentum to reach the speed of light, or, to put it another way: it can't be done. As we push anything faster, it becomes increasingly difficult to push it still faster. This is sometimes summarised as 'the faster you go, the heavier you get'.

However fast we wanted to get our train carriage going, whether that's a realistic 90 miles per hour, or a ludicrous, 90% of the speed of light, we're going to need energy to do so. If we have a steam train, that energy might come from burning coal. If we have a more modern train, that energy might come from electrified rails. If our

SATURN.

train's broken, and we have to push it up to speed ourselves, then the energy will come from the food that we've eaten.

If we have a look at how momentum varies with speed in Figure 6.3, we might hazard a guess that, to get our train running at relativistic speeds, we're going to need more energy that we would've expected from Newton's model of momentum. It was calculating exactly how much energy would be required to do so that became Einstein's most famous equation.

CHAPTER 7

Work in Progress

IN RELATIVITY, EVERYTHING stems from the fact that the speed of light is an absolute constant. To maintain this constant, we've seen how relativity affects not only how things are observed, but directly affects how things behave. We now know how to calculate the momentum of our train as it speeds past the platform (of course, we'd need to know the precise mass of the train and how fast it was travelling). But how could we get the train up to speed in the first place? To do that, we're going to need some energy.

'Energy' is one of those terms that we often hear in everyday parlance. We might describe a piece of music or a lively puppy as 'energetic'. In our context, 'energy' has a very specific meaning. The best way to understand precisely what we mean is with an example: If our train is stationary by the platform, exactly how much of this 'energy' do we need to get our train going up to a particular speed? That speed could be 90 miles per hour, or 90% of the speed of light.

It's important to learn how to walk before we learn how to run. Before we think about how much energy we'd need to have our train running at a 'relativistic speed', let's start by thinking about how much energy we'd need to have our train chugging along at a more reasonable, 'non-relativistic' speed. This way, we don't have to concern ourselves with the effects of the Lorentz factor, which considerably simplifies our situation. Although this simplification will limit our result to non-relativistic speeds, it will still be a useful and instructive result, valid for the overwhelming majority of everyday objects which travel at non-relativistic speeds. Once we have our solution for non-relativistic speeds, we'll be good and ready to replace the Lorentz factor and see how much more energy we need to reach truly relativistic speeds.

We're going to get our train moving by pushing it along the track with a constant force. As we push the train along, it's going to get faster and faster, until we reach our target speed, and we can stop pushing. Once the train is up to speed, the energy that the train has due to its motion, its 'kinetic energy', will be equal to the amount of work that we had to exert in pushing the train along. The amount of work that we have to do depends on two things: how hard we push the train, and how far we push it. If we push twice as hard, we only have to push it half as far. We'll do the same amount of work either way. If we're pushing the train with a constant force, then the amount of work we need to do is the force that we're pushing with, F, multiplied by the distance that we need to push the train for, x:

$$W = Fx.$$

We can see this relationship illustrated in Figure 7.1. But how much force do we need to push with, and how far will we need to push? The only thing we know for sure is that we need to get our train up to a speed of 'V'.

If our velocity was constant, we could calculate how far we would need to push with the relationship that velocity is equal to distance divided by time, so distance is equal to velocity multiplied by time. This relationship is shown in Figure 7.2.

However, if we push our train with a constant force, the velocity won't be constant; instead the *acceleration* will be constant. Newton's Second Law of motion tells us that force is equal to mass times acceleration, so we can write down the amount of work that we need to do in terms of the mass and acceleration of the train:

$$W = (m\,a)\,x.$$

How can we relate the distance and the acceleration to the final speed?

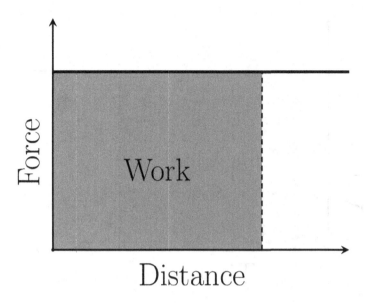

Figure 7.1: This is a graph of 'Force vs. Distance'. It could represent, say, pushing a book across a desk, or pushing any object with a constant force. If the force is constant, we can find out how much 'Work' is done by multiplying the force by the distance. In terms of the graph, this corresponds to the shaded grey box; this area corresponds to 'Force times Distance'.

Acceleration is the rate of change of velocity: how much our velocity has changed in a certain amount of time. For example, a high-performance sports car might be able to accelerate from zero to sixty miles per hour in a time of three seconds. We could say that this car accelerates at a rate of twenty miles per hour per second. As for our train, it's going to accelerate from zero to a speed of V. Since we don't know how long it's going to take us to push the train up to this speed, let's just call this time t. That way, we can relate our acceleration to our target velocity: $a = V/t$.

Now we can relate the amount of work that we'll need to exert with out target velocity, and the time that it's going to take for us to do that much work:

$$W = m \left(\frac{V}{t}\right) x.$$

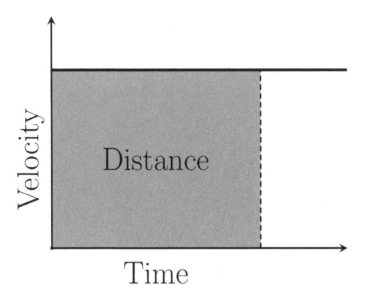

Figure 7.2: This graph might look similar to the graph of 'Force vs. Distance', but it's actually quite different. We now have Velocity on the vertical axis and Time on the Horizontal axis. This particular graph represents an object moving with a constant velocity, for example, a car driving with a constant speed on a motorway. For a constant velocity, if we multiply it by how long our car has been driving, we'll find the total distance travelled. On this graph, that distance corresponds to the 'area under the graph' – the box shaded in grey.

We still don't know how far we'll have to push the train, x, or how long it's going to take us, t. But we do know that our final velocity is going to be V. How far are we going to have to push? If we say that we're going to push for a length of time of t, how far will our train go in this time?

If we were moving with a constant speed, let's say u, then in a time of t we'd travel a distance of ut. But the whole point of pushing the train is to change our velocity, from zero up to V, so our speed isn't constant. We can see this situation illustrated in Figure 7.3. In our situation the distance that we'll cover, x, will be

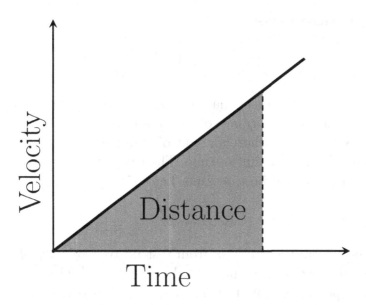

Figure 7.3: This is still a graph of of 'Velocity vs. Time', but – as we can see – now our hypothetical object starts from a velocity of zero, and then steadily increases speed as time goes by. This might correspond to a train as it's pulling away from a station. Unlike before, we can't now calculate the Distance travelled just by multiplying the Velocity and the Time, because the Velocity is constantly changing. However, the principle still holds that the Distance is the 'area under the graph' – shaded in grey. This area isn't always straightforward to calculate, but in this particular case of the velocity constantly increasing, the area is a triangle, and we can calculate the area using geometry.

the same as the distance that we'd travel if we pushed for the same length of time, t, at our average speed during this time. So what's the average speed of the train while we're pushing it form zero up to a speed of V? Since we're pushing with a constant force, the average speed during this time is going to be $1/2\,V$. So the distance that we're going to push for is going to be $x = 1/2\,Vt$. Let's substitute this in for our equation for the amount of work that we're going to need to do:

$$W = m \left(\frac{V}{t}\right) \left(\frac{1}{2} V t\right).$$

Something very interesting happens once we've substituted this in: the two values of t are going to cancel out. We can either push the train very hard for a short amount of time, or push the train less hard, for a longer amount of time. The total amount of work that we'll need to do will be the same. Let's simplify our equation for the work:

$$W = \frac{1}{2} m V^2.$$

So to push a train of mass m from rest up to a velocity of V, the amount of work that we need to do is $1/2 \; m V^2$. Once the train is up to speed, we can stop pushing. In the absence of friction, the train will continue with a velocity of V. The energy that we used in exerting a force over a distance has been transferred to the 'kinetic energy' of the train:

$$E_K = \frac{1}{2} m V^2.$$

This equation may be familiar to anyone who has already studied a little physics. This equation tells us how much energy is required to get an object with a mass of m moving with a speed of V. It tells us that if we want to double our speed, we require quadruple the energy.

If we simply took this equation at face value, we might be tempted to calculate how much energy we would need to get our train moving at the speed of light. However, we couldn't use this equation to calculate how much energy we would need to get our train carriage going along at relativistic speeds, because we found this relationship using Newton's Second Law of Motion, $F = ma$.

In the previous chapter, we saw that, at relativistic speeds, Newton's relationship between mass, speed, and momentum has to be modified to hold for relativistic speeds. Instead of just $p = mv$, we need to include the Lorentz factor, so that we have $p = \gamma m v$. To calculate how much energy we would need to get our train going along at relativistic speeds, we need a similarly updated version of Newton's Second Law of Motion, which holds for relativistic speeds.

We'll see how to do this in the next chapter. Before we do so, there's another question that requires our attention. In our example here, we were able to calculate our kinetic energy just with arithmetic because, in Newton's Second Law of Motion, force only depends on an object's mass and acceleration, not its speed. If the force also depended on speed, we wouldn't be able to use this approach. How could we calculate how much work we'd need to do then?

What if we only considered doing a sufficiently tiny amount of work on the train that the force is approximately constant while we're pushing it? We'd only change the speed of the train by a correspondingly tiny amount. To calculate the total amount of work we'd need to do, we would need to add up all of these tiny amounts of work, from the train sitting stationary at the platform, until it's going along at speed down the track. How could we go about adding up all these tiny individual bits of work?

In addition to devising his Laws of Motion, Isaac Newton (and his contemporary, Gottfried Leibniz), devised a very elegant way to do exactly that. The method is called 'integration', and it allows us to exactly add up the net amount of lots of infinitesimally small things. Anyone who is a little unfamiliar with integration will discover a wealth of explanation to the concepts in any introductory reference on calculus. For our purposes here, we will only need a couple of key results, which we'll introduce as we come to them.

Before we jump back to relativity, let's see how we can use this approach to reproduce our result for how much work we need to do to get our train going along at non-relativistic speeds. By seeing how the calculation works with non-relativistic physics, we'll be better prepared to understand how it works in the relativistic case.

Let's start by thinking about pushing the train just a very small distance, dx. What does the 'd' in front of the x mean? The little d means that we're only going to push for sufficiently small distance that we don't have to worry about any changes in the force that we push with.

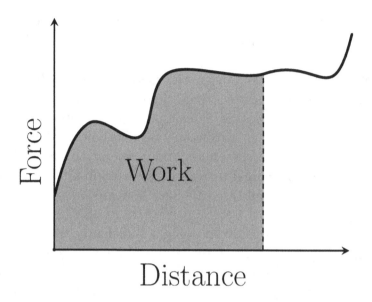

Figure 7.4: This is another graph of 'Force vs. Distance', but now the Force isn't constant as we're pushing our hypothetical object. We can't calculate the Work Done by multiplying the Force by the Distance, because the Force isn't constant. However, the principle still applies that 'Force is the area under the curve', and we can calculate this area using integration.

If we only exert this force over a tiny distance, then we're only going to do a correspondingly tiny amount of work: $dW = F\,dx$. To get the train up to speed, we need to add up all of these tiny little bits of work, starting from a speed of zero, and pushing until we reach our target speed of V. That will tell us the total amount of work that we will need to do. We can start writing this down mathematically like this:

$$E_K = \int_0^V dW$$

This relationship is illustrated in Figure 7.4. That integration sign is actually a very stretched out letter 'S', and tells us to 'Sum' all of those tiny amounts of work, dW, starting from a speed of zero, until we get to a speed of V. How to we go about this?

Let's remember that to push the train a small distance, dx, we have to do an amount of work given by $dW = F\,dx$. Let's substitute that relationship into our integral:

$$E_K = \int_0^V F\,dx$$

This is a very important relationship, so we've put a box around it, because it will be very useful for us later.

The limits of our integration are the initial and final velocities of our train, but it's not immediately apparent how the quantities that we're integrating, $F\,dx$, depend on velocity. What we need to do is re-write $F\,dx$ in terms of velocities. The key to making this work is linking the force, F, and the distance, dx, by thinking about how they both relate to momentum.

If we exert a force F over a small distance dx, then we'll do a small amount of work, dW. What if we think about exerting a force for a small amount of time, dt? If we push with a force of F for a time of dt, then we'll change the momentum of the train by a small amount, $dp = F\,dt$. With a little bit of re-arranging, we can see that our force is equal to that small change in momentum, divided by the small amount of time:

$$F = \frac{dp}{dt}.$$

This is also a very important relationship for later. What will a small change in momentum look like? Remember that in this example, we're just thinking about Newtonian momentum, when we're not moving with relativistic speeds, so our relationship between momentum and speed is:

$$p = mv$$

So our force will be equal to a small change in mv, divided by a small amount of time, dt:

$$F = \frac{d(mv)}{dt}$$

We're keeping the mass of our train the same, so the only way we can change the momentum is by changing the velocity, so our force must be

$$F = m \frac{dv}{dt}.$$

This relationship for Force is just another way of writing Newton's Second Law, $F = ma$, since acceleration is the rate of change of velocity. Let's substitute this equation into our integral:

$$E_K = \int_0^V m \frac{dv}{dt} \, dx.$$

What if we just re-arrange how all the terms are written? We could have:

$$E_K = \int_0^V m \frac{dx}{dt} \, dv.$$

Let's be clear that this velocity *little-v* isn't the same as the final velocity, *Capital-V*. *Capital-V* is the final velocity that we're trying to reach. *little-v* represents the intermediate velocity of the train while we're accelerating up to *Capital-V*. Let's also remember that term dx/dt is just the velocity of the train, *little-v*. Substituting this in to our integral, we find:

$$E_K = \int_0^V mv \, dv.$$

We now have an integral in terms of velocities, and we're ready to add up all of those little parts. Since the mass of our train isn't going to change, we can factor it out in front of the integral:

$$E_K = m \int_0^V v \, dv.$$

So how do we add up all of those tiny bits? It's here that we're going to quote our first result using integration. Whenever we have an integral of the form:

$$Y = \int_a^b X \, dX$$

the result is:

$$Y = \left[\frac{1}{2} X^2 \right]_a^b.$$

In this example, a, b, X and Y are just arbitrary variables. What do we do about the a and b on the right-hand-side of the square brackets? It's just a shorthand notation to subtract the quantity inside the square brackets at the upper value from the same quantity at the lower value, so we have:

$$Y = \frac{1}{2}b^2 - \frac{1}{2}a^2$$

This is exactly the solution that we need. Let's remind ourselves of the integral that we're trying to solve:

$$E_K = m \int_0^V v \, dv$$

We can now use our standard result, and find:

$$E_K = m \left[\frac{1}{2}v^2 \right]_0^V$$

Let's put in the limits of the integration:

$$E_K = m \left(\frac{1}{2}V^2 - \frac{1}{2}0^2 \right)$$

and now with just a little simplification, we find exactly the same result as before:

$$E_K = \frac{1}{2}mV^2$$

This might all seem like a great deal more mathematics only to arrive at the same conclusion as before, but this approach affords us a great deal more flexibility. For example, in a real-world situation, the locomotive force of the engine may vary with the speed of the train. On the other hand, there may be opposing forces of friction and air resistance, which increase as the train goes faster. In our first calculation of the kinetic energy of the train, we would have no way to include these effects. With the approach of using integration, as long as we can write down how the force depends on the speed, we can calculate the energy required.

Turning our attention back to relativity, we aren't going to concern ourselves with any confounding business such as the mechanical output of the locomotive, or the effects of air resistance or

66 ■ First Steps in Space-Time

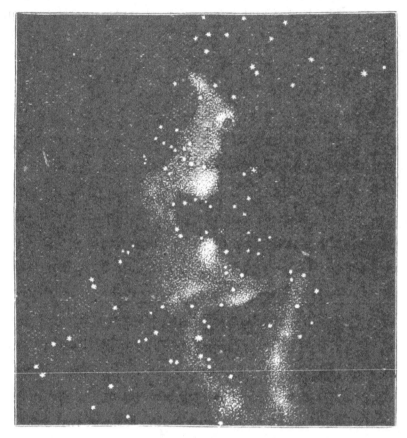

PART OF MILKY WAY.

suchlike. However, as we'll soon see, the apparent force exerted on the train will depend on its speed. With the work and energy that we've invested in understanding the calculations in this chapter, we're now ready to extend our approach to relativistic speeds.

CHAPTER 8

A Momentous Integration

IN OUR EXPLORATION of special relativity, we've seen how relativity affects how events are observed and speeds are combined, and also how momentum is affected by relativity. We've seen how, with relativistic momentum, objects effectively become 'heavier' as their speed increases. In the previous chapter, we briefly removed our relativity hat to focus on the relationship between momentum, force and energy.

We've seen that, in order to change the momentum of our train, we need to apply a force. We've seen that, to apply a force over any distance, we need energy. We calculated exactly how much energy we would need to get our train moving from a standstill, up to a speed that we called V. We discovered the 'kinetic energy' of the train, the energy the train has due to its motion, given by:

$$E_K = \frac{1}{2}mV^2$$

We first found this result by thinking about how much work we would do if we exerted a constant force on the train over the requisite distance to bring it up to speed. We then repeated the calculation with an integral, which would allow us to account for a force which varies with speed. The integral we calculated was:

$$\boxed{E_K = \int_0^V F\, dx}$$

which is just a mathematical way of stating that, to apply a force over a distance, we need energy. We also had to examine the relationship between force and momentum:

$$\boxed{F = \frac{dp}{dt}}$$

This equation is just a mathematical way of stating that, to change our momentum, we need to apply a force for a period of time.

The kinetic energy that we calculated in the previous chapter is only valid for non-relativistic speeds, because we used the non-relativistic relationship between momentum and velocity: $p = mv$. While this relationship is only valid for non-relativistic speeds, the general relationships that force is the rate of change of momentum, and that work is the integral of force over a distance, aren't limited to non-relativistic speeds. Both of those boxed equations are valid for any speed, up to the speed of light.

It's now time to put our relativity hat back on. What if we repeated our derivation from the previous chapter, but with the relativistic relationship between momentum and velocity? Then we could calculate how much energy we would need to get our train going at any speed we like. That's the plan.

Let's remember that relativistic momentum has a Lorentz factor, γ:

$$p = \gamma m v$$

Even though our expression for momentum is different, force is still the rate of change of momentum with respect to time:

$$F = \frac{dp}{dt}$$

Let's now substitute in our expression for relativistic momentum:

$$F = \frac{d(\gamma m v)}{dt}$$

We now have the beginnings of a relativistic equation for force. In the non-relativistic case, when we had $p = mv$, this approach gave us Newton's famous Second Law of Motion, $F = ma$, (because the rate of change of velocity with respect to time is acceleration). Does this mean that our relativistic force will just be:

$$F = \gamma m a \,?$$

Let's not forget that the Lorentz factor also depends on velocity:

$$\gamma = \frac{1}{\sqrt{1 - \left(\frac{v}{c}\right)^2}}$$

So, in full, our relativistic force equation is:

$$F = \frac{d\left(\frac{mv}{\sqrt{1-\left(\frac{v}{c}\right)^2}}\right)}{dt}$$

Anyone who's familiar with a little calculus might suspect that this is going to make our relativistic force equation a bit more complicated. We're going to save ourselves a page or two of mathematics here by leaving the evaluation of this derivative to the more enthusiastic students of calculus. The good news is that taking the derivative is essentially a mechanical process, and the end result is rather elegant:

$$\boxed{F = \gamma^3 m a}$$

Whereas in Newtonian physics we have that $F = ma$, the relativistic equivalent has an additional factor of γ^3.

Let's remember that acceleration is the rate of change of velocity with respect to time, $a = dv/dt$, so we could also write:

$$F = \gamma^3 m \frac{dv}{dt}$$

We now have an expression for the force, a generalisation of Newton's Second Law of Motion, which is valid for relativistic speeds. We're now ready to substitute this relationship for the force into our integral for the kinetic energy:

$$E_K = \int_0^V F \, dx$$

Substituting in our relativistic force, we find this integral:

$$E_K = \int_0^V \gamma^3 m \frac{dv}{dt} dx$$

As with the classical example, we can now do a little bit of rearranging, and have an integral with respect to velocity, instead of position:

$$E_K = \int_0^V \gamma^3 m \frac{dx}{dt} dv$$

And, exactly as before, we can note that $dx/dt = v$, so our integral becomes:

$$E_K = \int_0^V \gamma^3 m v \, dv$$

This looks exactly like the integral in the classical version, except for the extra factor of γ^3, because of our relativistic force. If we were being hasty, we might conclude that this means our kinetic energy will simply have an additional factor of γ^3, and be $1/2 \, m \, v^2 \, \gamma^3$. But, as before, let's not forget that γ depends on v. Let's expand our integral out, to see what it looks like in full:

$$E_K = \int_0^V \frac{mv}{\sqrt{1 - \frac{v^2}{c^2}}^3} dv$$

That extra factor of γ^3 has made the whole situation a little more complicated.

Solving this integral is essentially a mechanical process, albeit one that would consume several pages of equations. As such, in the same manner as before, we're going to skip over the mechanics of the integration, and focus directly on the end result. After evaluating the integral, we find that our kinetic energy is:

$$E_K = \left[\frac{mc^2}{\sqrt{1 - \frac{v^2}{c^2}}} \right]_0^V$$

We might notice that this result also contains the Lorentz factor, so we can simplify the result to:

$$E_K = \left[\gamma m c^2 \right]_0^V$$

All we have to do now is substitute in the limits of the integration: 0 and V. Substituting in the upper limit of integration will give us:
$$\gamma mc^2$$
The interesting thing to note is with the lower integration limit. For $v = 0$, γ doesn't equal zero, it equals 1. We'll have terms from both the upper and lower integration limits, so our relativistic kinetic energy is
$$E_K = \gamma mc^2 - mc^2$$
Let's rearrange this so that we have:
$$\gamma mc^2 = E_K + mc^2$$
Let's call γmc^2 the 'total energy':
$$\boxed{E_T = \gamma mc^2}$$
The important result here is that our total energy is comprised of our kinetic energy, plus some extra energy, which we still have, even if our kinetic energy is zero:
$$\text{If } E_K = 0 \text{ then } E_T = mc^2$$
Let's call this our 'rest energy'; the energy we have, even if we're at rest.

We can see how our relativistic kinetic energy compares to the Newtonian equivalent in Figure 8.1. For speeds towards the lower end of the scale, we can hardly see the difference between the two models. As we increase speed, the Newtonian model only increases steadily as we approach light speed, and predicts that we would need a large, but finite, amount of energy to reach light speed. On the other hand, our relativistic kinetic energy shoots up dramatically as our speed increases, and tends to infinity as our speed approaches the speed of light. To reach light speed, we would need an infinite amount of energy!

It's very instructive to see how our total energy is related to our rest energy and our kinetic energy when we're at non-relativistic

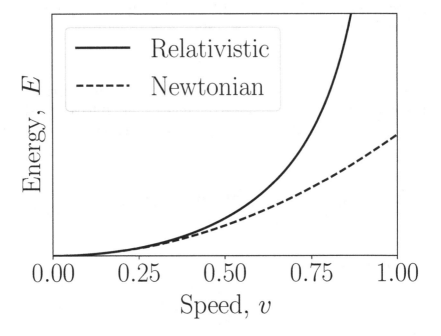

Figure 8.1: In this plot, we can see how kinetic energy, E depends on speed, v. In the classical model, energy only increases fairly steadily with speed, whereas in the relativistic model, energy dramatically shoots up as we approach the speed of light. To reach the speed of light, we would need an infinite amount of energy.

speeds. We can use this handy mathematical approximation, which is valid whenever an (arbitrary) variable, X, is small:

$$\frac{1}{\sqrt{1-X}} \approx 1 + \frac{1}{2}X$$

The wavy equals sign tells us that the relationship is only 'approximately equal'. So, when our velocity v is small compared to c, we can approximate the Lorentz factor as:

$$\frac{1}{\sqrt{1-\frac{v^2}{c^2}}} \approx 1 + \frac{1}{2}\frac{v^2}{c^2}$$

Let's remember that our total relativistic energy is:

$$E_T = \frac{mc^2}{\sqrt{1 - \frac{v^2}{c^2}}}$$

For small speeds, v, we can now approximate this energy as:

$$E_T \approx \left(1 + \frac{1}{2}\frac{v^2}{c^2}\right)mc^2$$

Let's factor through the mc^2:

$$E_T \approx mc^2 + \frac{1}{2}mv^2$$

We find that, for non-relativistic speeds, our total energy is our rest energy, plus the classical kinetic energy that we found in the previous chapter. If we are stationary, our kinetic energy is zero, and classical physics would tell us that our total energy is also zero. However, relativity tells us that, even if our kinetic energy is zero, we still have some 'rest energy', just because we have mass, given by:

$$\boxed{E = mc^2}$$

The concept of 'rest energy' was one of the most extraordinary results from relativity. This equation tells us that even when an object isn't moving, it still has some energy, just by virtue of its mass.

Although it was $E = mc^2$ that became the defining equation of the twentieth century, Einstein originally wrote the relationship down as:

$$m = \frac{E}{c^2}$$

which evidently proved nowhere near as catchy as the exactly equivalent $E = mc^2$.

What does this tell us? It tells us that our train carriage has a greater mass when it's moving, because of its kinetic energy. It tells us that a fully charged battery weighs more than a drained battery, because of the electrical potential energy it contains. It tells us that a hot cup of coffee has more mass than the same coffee when it's cooled down, because of the thermal energy when the coffee is hot. For any remotely reasonable cup of coffee, the mass of its thermal energy is utterly negligible, and our cup of coffee is going to lose far more mass as it cools due to evaporation.

Fortunately for us, matter generally takes considerable persuasion to transform itself into energy. However, for a heavy atomic nucleus, a significant percentage of the mass is due to the energy required to bring the constituent protons and neutrons sufficiently close together to form the nucleus. It's this energy which is released in a nuclear reactor, or to devastating effect in a nuclear weapon. As for the protons and neutrons themselves, their mass is largely due to the energy which binds their constituent particles, the 'up' and 'down' quarks. Through $E = mc^2$, Einstein came to realise that the mass of a particle is not an arbitrary intrinsic property, but a reflection of the energy required to assemble it.

Everyone has heard of $E = mc^2$. Most people known that the E stands for 'Energy', the m stands for 'mass', and the c stands for 'the speed of light'. If there's one thing that people know about the speed of light, they know that its speed is a constant. Of these, some know that $E = mc^2$ arises as a consequence of this fact. We've now seen exactly why this is so, and in the next chapter, we'll reflect on why it's so important.

DOUBLE SPIRAL NEBULA.

CHAPTER 9

Time to Reflect

WE HAVE COME A LONG WAY since the first tick of Alice's light clock. To start with, our humble objective was to make the speed of light a constant not just for Alice, but also for Bob. Once we started pulling at the thread of time as an absolute constant, the whole ball of absolute notions of length, momentum, force, and energy unravelled. After arriving at the conclusion that matter is another form of energy, we might ask ourselves: how important is relativity to everyday life? Relativity predicts that whenever we're moving relative to anyone else, they observe us travelling through time at a slower rate. Let's think about the most extreme effect this might have on an actual person.

The fastest trains in the world travel at about one-hundred metres per second. What if Alice spent her whole career as a high-speed train driver? Let's imagine she spends fifty hours per week driving trains, for fifty weeks of the year, over a fifty year career. That's a total time of 125,000 hours, or 450 million seconds. If Alice spent this entire time driving a train at 100 metres per second, how much younger would she be than Bob, who spent his whole career stationary?

To work this out, we need to calculate the Lorentz factor, for a speed of 100 m/s:

$$t = \frac{t'}{\sqrt{1 - \left(\frac{v}{c}\right)^2}}.$$

The problem is that the ratio of v/c is so small, we simply can't calculate this on a calculator. However, whenever that ratio is really small, we can use this handy mathematical relationship:

$$\frac{1}{\sqrt{1-X}} \approx 1 + \frac{1}{2}X.$$

As we've seen before, this is a general mathematical relationship, where X is just a variable. With this relationship, we find that, when v is small, our times can be related by:

$$t \approx t'\left(1 + \frac{1}{2}\frac{v^2}{c^2}\right).$$

Using this approximation allows us to focus on the difference between t and t', which we can now calculate directly. We find that after fifty years driving high speed trains, Alice is 25 microseconds younger than she would've been had she taken a steady desk job, like Bob. Even if Alice was travelling much faster, there's still no escape from the arrow of time. So we still might fairly ask, how important is relativity?

We might be tempted to conclude that relativity is just an academic novelty, but without relativity, there would be no $E = mc^2$, and without $E = mc^2$, there would be no nuclear power. Via both civilian and military nuclear power, relativity has had a huge effect on the lives of everyone who lived in the second half of the twentieth century.

Beyond unleashing nuclear power on the world, $E = mc^2$ also finally provided an answer to a long-standing cosmic mystery. For decades, physicists and astronomers had been increasingly perplexed by the utterly extraordinary volumes of power emanating from the Sun and the stars. There was simply no mechanism which could explain how the Sun and the stars could output so much power, for so long.

The first ever nuclear bomb was powerful enough to be felt over a hundred miles away, and hot enough to melt the desert sand into glass. However, this vast explosion was caused by transforming just one gram of matter into energy – about the mass of a few grains of rice. In the Sun, four thousand times as much mass is transformed into energy, every nanosecond. Without $E = mc^2$, we would be utterly at as loss to explain how the Sun and all the other stars could output such stupendous amounts of power, for billions of years. Beyond $E = mc^2$, there's another vital result of relativity, which affects the lives of everyone on the planet, on a daily basis.

Let's consider a scenario with a solitary electron. Close to this electron is a copper wire, with many more electrons flowing through it. Even though the electrons in the wire are moving, there's exactly as many stationary protons in the copper of the wire. Overall, (like the old joke about the neutron who walks into a bar), there's no charge. Outside the wire, our solitary electron is just sitting there, not seeing any net positive or negative electric charge. It's going to be quite content just sitting there. But what if we get that little electron moving along?

Let's imagine we get the electron moving with the exact same speed as the electrons in the wire. To our little electron outside the wire, it looks like the other electrons in the wire are at rest, and now the protons are the ones moving along (but in the opposite direction). But if the protons are moving, they're going to appear length-contracted. To the little electron outside the wire, it's going to look like there are now more protons than electrons. What effect is this going to have on the little electron?

The electron's now going to see a net positive charge on the wire. What effect is this going to have? If it's moving along next to the wire, our little electron is going to feel an attractive force towards the wire. It's going to start spiralling in a corkscrew as it travels along the wire. Let's imagine we bring the little electron to a stop. Now it looks like the wire doesn't have a net charge anymore. Like a ghost, the force has disappeared, only to reappear again if the electron ever starts moving.

This transient effect is called the 'Electromagnetic Force', and solves one of the last great puzzles of classical physics. This force had long been known to the pioneers of electromagnetism, but, before relativity, the explanation had remained a mystery. Relativity provided the last piece of the puzzle of classical electromagnetism.

When we think of electromagnetism, we might picture a giant electromagnet in a scrapyard picking up the wrecked remains of a vehicle. But electromagnetism has proved to be of far wider value than just to the metal scrappage industry. Understanding electromagnetism is essential for electric motors and generators, and radio transmitters and receivers. Without relativity, our theoretical understanding of these effects (and the myriad contraptions which depend on them) would be incomplete.

Relativity made Einstein the defining, archetypal image of a scientist for generations. However, it's important to remember that the work stands on the shoulders of centuries of work by hundreds of others. Our journey began on a foundation of mathematics from Pythagoras, and has been shaped throughout by the work of Galileo, Kepler, Leibniz and Newton.

Our story began in earnest with Michelson's confident prediction that 'the future of Physical Science has no marvels in store even more astonishing than those of the past'. This prediction was in part due to the extraordinary success of Maxwell's electromagnetism, which unified the work of Ampère, Ohm, Gauss, Henry and Faraday, and predicted a constant speed of light.

We should remember the careful observations of Newcombe, Michelson, and Morley, and the theoretical ideas of Voigt, Fitzgerald and Larmor, which all set the stage for relativity.

We've studied the elegant framework of Lorentz, whose mathematical transformations are the foundation of special relativity. We've seen how Einstein interpreted Lorentz's results, and saw the implications for momentum and energy.

A century of relativity culminated with the famous detection of 'gravitational waves', almost exactly one hundred years after Einstein first proposed the general theory of relativity. In a wonderful twist of symmetry, the vast devices which detected the gravitational waves were in fact stupendously enormous versions of the interferometer first designed by Michelson to measure the speed of light, which sparked relativity in the first place. It's hard to image what Michelson would've made of everything that relativity held in store. Hopefully, he would've thought it was all marvellous.

THE EARTH SEEN FROM THE MOON.

FURTHER READING

Basset, Bruce (2002), 'Introducing Relativity: A Graphic Guide', *Icon Books Ltd*, ISBN: ISBN: 1840463724

Einstein, Albert (1916), 'Relativity: The Special and the General Theory', *New York: Three Rivers Press*, ISBN: 978-0-517-88441-6

Ferreira, Pedro G. (2014), 'The Perfect Theory: A Century of Geniuses and the Battle Over General Relativity', *Little, Brown*, ISBN: 978-1-4087-0430-1

Lambourne, Robert (2010), 'Relativity, Gravitation and Cosmology', *Cambridge University Press*, ISBN: 978-0-5211-3138-4

Lieber, Lillian, and Lieber, Hugh Gray (1945). 'The Einstein Theory of Relativity: A Trip to the Fourth Dimension', *Paul Dry Books*, Inc, ISBN: 978-1-5898-8044-3

Bibliography

[1] Bruce Bassett and Ralph Edney. *Introducing Relativity*. Introducing series. Icon Books, Basingstoke, England, October 2002.

[2] Frank Watson Dyson, Arthur Stanley Eddington, and C. Davidson. Ix. a determination of the deflection of light by the sun's gravitational field, from observations made at the total eclipse of May 29, 1919. *Philosophical Transactions of the Royal Society of London. Series A, Containing Papers of a Mathematical or Physical Character*, 220(571–581):291–333, 1920.

[3] A. Einstein. Ist die trägheit eines körpers von seinem energieinhalt abhängig? *Annalen der Physik*, 323(13):639–641, 1905.

[4] A. Einstein. Zur Elektrodynamik bewegter Körper. *Annalen der Physik*, 322(10):891–921, January 1905.

[5] Pedro G Ferreira. *The perfect theory*. Little, Brown, London, England, February 2014.

[6] Stephen Hawking. *Brief history of time*. Econo-Clad Books, Topeka, KS, October 1999.

[7] H. Lorentz. Electromagnetic phenomena in a system moving with any velocity smaller than that of light. *Proceedings of the Royal Netherlands Academy of Arts and Sciences*, 6:809–831, 1904.

[8] Albert A. Michelson and Edward W. Morley. On the Relative Motion of the Earth and of the Luminiferous Ether. *Sidereal Messenger*, 6:306–310, November 1887.

Index

A
Acceleration, 63
Ampère, André-Marie, 7, 86

B
Braque, Georges, 1

C
Cubism, 1–2

D
Donati's comet, *47*
Double Spiral Nebula, *81*

E
Earth seen from moon, *87*
Eddington, Arthur, 9
Einstein, Albert, 2, 7–10, 49, 57–58, 80, 86
Electrical circuits, 7
Electromagnetism, 2, 7, 85
Energy, 59

F
Faraday, Michael, 7, 86
Fitzgerald, George, 6, 86
Force, 60, 74

G
Galilei, Galileo, 42, 86
Gauss, Carl Friedrich, 7, 86
General relatively, 8
Gleizes, Albert, 1
Gravitational waves, 86

H
Henry, Joseph, 7, 86

I
Inverse Lorentz transformation, 36–37

K
Kepler, Johannes, 4, 86
Kinetic energy, 60, 65, 73–74, 76

L
Larmor, Joseph, 6, 86
Leibniz, Gottfried, 66, 86
Length measurement, 26
Lorentz, Hendrik, 2, 6, 19
Lorentz factor, 19, 21–22, 27–30, 33, 56, 59–60, 65, 74–76, 79, 83

Lorentz transformations, 36–37, 43
   for time, 37
   for time and position, 38

M
Maxwell, James Clerk, 7, 86
Mercury, 4
Metzinger, Jean, 1
Michelson, Albert, 2, 4, 6, 86
Michelson Interferometer, *5*
Michelson & Morely experiment, 6–7
Milky Way galaxy, 7, *71*
Minkowski, Hermann, 8
Momentum, 49, 51–52, 54, 68, 74
   conservation equation, 55
   defined, 52
Morely, Edward, 5–6, 86

N
Neutron, 85
Newcombe, Simon, 4, 86
Newton, Isaac, 49, 86
Newtonian mechanics, 2
Newtonian model, 77
Newton's law of motion, 66
Newton's model of momentum, 53, 57, *57,* 58
Newton's second law of motion, 60, 65–66, 68–69, 74–75, 80, 84–85
Non-relativistic speeds, 59, 66

O
Ohm, Georg, 7, 86
Orbit, 4

P
Photons, 12, 15, 42, 44
Picasso, Pablo, 1
Planck, Max, 8
Poincaré, Henri, 2, 8
Pythagoras, 86
Pythagoras's theorem, 16

R
Relative motion of the Earth, 6
Relativistic energy, 79
Relativistic force equation, 75–76
Relativistic momentum, 56–57, 73–74
Relativistic speed, 56, 59
Relativistic velocity, 56
Relativity, 2, 7, 9, 12, 49, 59
Rest energy, 77, 79
Riemann, Bernhard, 8

S
Saturn, *58*
Schwarzschild, Karl, 8
Simultaneous equations, 36
Solar eclipse, *10*
Special relativity, 8, 73
Speed, 16
Speed measurement, 25–26, 33–38
Speed of light, 12, 14, 25–26, 29, 45, 55

Sun
    in eclipse, *23*
    volcanic eruption, 38

T
Thermodynamics, 2
Time dilation, 19

V
Voigt, Woldemar, 6, 86

Printed in the United States
by Baker & Taylor Publisher Services